MW01067789

WHAT THE BODY COMMANDS

WHAT THE BODY COMMANDS

The Imperative Theory of Pain

Colin Klein

The MIT Press
Cambridge, Massachusetts
London, England

MIT Press books may be purchased at special quantity discounts for business or sales promotional use. For information, please email special_sales@mitpress.mit.edu.

This book was set in Times New Roman by the author, using LaTeX.
Printed and bound in the United States of America.

Library of Congress Cataloging-in-Publication Data

Klein, Colin, 1979–
What the body commands : the imperative theory of pain / Colin Klein.
p. cm
Includes bibliographical references and index.
ISBN 978-0-262-02970-4 (hardcover : alk. paper)
1. Senses and sensation. 2. Pain—Philosophy. I. Title.
BD214.K54 2015
152.1'824—dc23

2015009282

10 9 8 7 6 5 4 3 2 1

For Paul R.

Physical pain is thus the psychical adjunct of an imperative protective reflex.
–Sherrington, The Integrative Action of the Nervous System.

In pain I hear the captain's command: "Take in the sails!"
–Nietzsche, The Gay Science *§318*

Contents

Acknowledgments

When I was thirteen, Dan Matyniak fractured my ankle in wrestling practice. A sympathetic E.R. doctor pumped me full of morphine shortly after. The two events left me with a lifelong interest in pain, along with a dodgy ankle that provided plenty of examples. When I later came across references to morphine pain in college and remembered my own experiences, Bennett Helm encouraged my early interest in the topic. Sean Kelly helped me turn my initial ideas into something that looked like professional philosophy, and Tim Bayne was especially generous and encouraging when I went to submit my first article. On a hike around Howth Head, Chris Mole suggested that I take a few months and write a short book to develop my theory further. Rather than heed his good advice, I've taken a few years to write a long one. David Bain, Michael Brady, Jennifer Corns, and The Pain Project/Value of Suffering Project out of the University of Glasgow have all provided fruitful intellectual exchange over the past several years. Bain in particular has been the sort of opponent that one always hopes for, cheerfully pointing out inadequacies in my account in helpful ways. I've also benefitted especially from exchanges with Murat Aydede, Derek Baker, Anne Eaton, Joe Gottlieb, Dylan Olson, Marya Schechtman, Thom Park, Adam Pautz, Geoffrey Pynn, Kristen Stubbs, Rachael Zuckert, two classes of students from the University of Illinois at Chicago in my fourth-year seminar on pain, and an army of anonymous reviewers on both this manuscript and other earlier papers. I also received helpful feedback from audiences of the Princeton Philosophical Society, the Association for the Scientific Study of Consciousness, University College Dublin, Northern Illinois University, Franklin and Marshall College, the University of Glasgow, and the Australian National University (ANU).

The bulk of this book was written while I was a visiting fellow at the ANU Centre for Consciousness. I am extraordinarily grateful to David Chalmers for giving me the opportunity to visit the ANU and participate in its unique intellectual culture. Much of the book has been shaped by discussions during teatimes, meals, and bushwalks. The material that now forms chapter 13 was presented at an ANU Philsoc and received numerous helpful comments. As I was worrying about reason and authority, Nic Southwood suggested that I look into Raz, and Seth Lazar and Massimo Renzo kindly answered my novice questions as I stumbled in unfamiliar waters. Laurie Walker of Canberra was the best landlord a man could hope for; he and his family kept my spirits up during marathon writing sessions and embodied everything good about Australia.

This book was finished in part with support from Australia Research Council grant FT140100422. Philip Laughlin at MIT press was admirably responsive and patient with my dithering. Ginny Crossman's help was much appreciated in the final stages of preparing the manuscript, as was that of my research assistant Peter Clutton.

Three people deserve special thanks. Manolo Martínez has been an enthusiastic and thoughtful co-author. He has pushed me hard to develop several aspects of what has followed, and the work we have done together I could not have done on my own. Esther Klein has listened to more conversations about pain than any spouse should have to endure, and her constant challenges have kept me honest. David Hilbert has been a wonderful friend and colleague. He read and commented on numerous early drafts of papers. His tireless enthusiasm kept me encouraged. More than anyone else, he shaped how I think about philosophy of mind and the job of an empirically engaged philosopher.

December 2014
Sydney
Canberra
Chicago

1 Puzzles about Pain

1.1 Imperativism

I have a pain in my ankle—a dull ache, present intermittently for the past few weeks. Before that, there was a hot, throbbing pain for a few days. And before *that*, there was a brief twinge as my ankle rolled on the pitch.

Despite these variations in quality, intensity, and duration, each of these sensations was a pain. In virtue of being pains, they have some common feature. They share that common feature with sensations I have felt in a wide variety of other situations: when splattered with hot grease, cut shaving, eating spicy Sichuan food, bringing my hand toward a campfire, wrenching my wrist, and on and on. That common feature distinguishes pains from other sensations like the sight of a cricket ball, the smell of woodsmoke, or my hunger for bacon.

In what follows, I will give an account of that common feature. I will argue that pains are *imperatives*.[1] They are sensations with a content, and that content is a *command* to protect a part of your body. Typically, protection will involve limiting activity: the ache in my ankle tells me to avoid putting weight on it, the throb of a burn to protect my skin, and so on. All pains have imperative content, and that imperative content is what distinguishes them as pains.

Call this *imperativism* about pain.[2] Imperativism offers a compelling story about two otherwise puzzling features of pain. First, pains are strongly and directly *motivating*. The pain in my ankle motivates me to avoid walking if I can and to walk carefully if I do. Pains motivate *directly*: when I feel pain, I don't need to deliberate further about whether I am moved by it. To be in pain just *is* to be motivated. Pains' motivation is similarly *unmediated*: my pain motivates me regardless of my desires, values, or other motivational states. Indeed, pain is usually at odds with my desires (I'd really *like* to walk around normally, and but for the pain I would). Finally, pains give us *reasons* to act rather than just pushing us around. I might have other, better reasons to walk, reasons that override the reasons my pain gives me not to. Those reasons must be in addition to, and stronger than, the reasons provided by pain. Even if I were to become convinced that my ankle were fine, and that walking on it posed no threat, my pain would *still* give me a strong reason not to walk.

1 I have argued this in several places over the years. See especially Klein (2007, 2010, 2012).

2 I believe the name "imperativism" is originally due to David Bain (2011).

Motivation is so tightly connected to pain that it's hard to imagine it absent. As Hall (2008) notes, if someone claimed to feel pain but be completely unmotivated by it, we would ordinarily doubt that they were in pain at all. Philosophers may be familiar with well-known pathological cases where precisely this seems to happen, of course. I'll deal with these in chapter 11. For now, note that such cases are at least *strange* and demand explanation. They fascinate precisely because ordinary life offers nothing like them: usually there is no gap between pain and motivation.

Second, pains are remarkably *uninformative*. Although the pain in my ankle motivates me, it gives me few clues as to why it's actually there. I know, of course, that I sprained my ankle, and that the sprain causes my pain. That is not information carried *by* the pain, however: I only know my ankle was sprained because I turned it a bit, and the next day it looked like an angry grapefruit. That's good evidence that I sprained it. But I could be wrong. My physician took time to rule out alternative causes, and that was not an absurd thing to do. When I can't infer causes, I often have no idea why I'm feeling pain: a mysterious backache, say, gives no indication *whatsoever* about what is causing it. Pains are uninformative partly because their potential causes are so diverse. Some pains are caused by *actual* tissue damage: I feel a pain when, and because, the hot grease hits my arm. Other pains are caused by merely *potential* damage: the hand as it approaches the fire, the needle pressed just shy of puncture, the muscle stretched to its limit. Note that potential damage need not be *imminent* damage. Sitting in an odd posture causes pain, but I'd have to do that repeatedly and for a long time before I did any harm.

Pains of *exertion* have an even more tenuous relation to potential damage. Running causes my thighs to burn, and bench-presses make my pectoral muscles ache. But I probably can't run far enough to injure my legs, and bench-pressing hurts at weights well shy of my maximum ability.

Finally, there are pains of *recuperation*, which are associated with *past* damage. For all I know, the actual damage to my ankle has mostly healed. It remains inflamed and sensitive. That is probably a good thing. Although healed, the ankle is still weak. I am in danger of re-injury if I use it too vigorously. Pains of recuperation form the vast bulk of the pains we feel. In chapter 3, following P.D. Wall, I'll suggest that pains of recuperation are also the biologically essential ones. Imperativism is crafted in part to place pains of recuperation front and center.

These two features of pain—its strong motivating power, on the one hand, and its uninformative nature, on the other—distinguish it from sensations in other modalities. The sound of a magpie doesn't motivate me on its own. I often hear a magpie and do nothing. When its cry does motivate, what it motivates depends in complex ways on my other mental states. Sometimes the sound of a magpie causes me to look up, other times to duck, and still others to reach for the tennis racket. What I do depends on what I want and what I believe. Ordinary sensations are also informative about the world. Under ordinary conditions, the sound of a magpie informs me that there's a magpie about. I might be wrong about that; the link is not infallible. But the whole *point* of auditory (or visual, tactile, etc.) sensations is to inform me about the world.

Hence the puzzle. Ordinary sensations inform but don't necessarily motivate. Pains motivate without informing. That is why pain is unusual. A good theory about pain's nature should remove some of that puzzlement. I offer imperativism as that story.

Parallels with ordinary language imperatives are useful, and I'll draw them frequently. Ordinary imperatives also motivate without informing. If I say to you, "Shut the door!", I do so to get you to act. If you accept me as a legitimate source of commands, then my imperative will motivate you and give you a reason to close the door. My imperative doesn't convey any information about why I want you to close the door, it doesn't tell you anything about what the world is like (except indirectly), and its function is not to inform you. That's in part because commands aren't truth-apt: my command may be many things, but it is neither true nor false. The same is true of pains. That means there is at least a *prima facie* parallel between pains and commands. The rest of this book will cash out that parallel and show the philosophical benefits of doing so.

What I have described is not the ordinary view of pain. The ordinary view, I take it, is that pains inform about some bad bodily condition. Pain *is* like vision or hearing, except that it tells you about bodily damage instead of properties in the world beyond. Because that sort of information doesn't seem like enough to get you moving, motivation is handled by an additional appended component—an affective state, a desire, an evaluation, or something along those lines. Theorists debate just what the two components are, what their ontological status is, and whether both are strictly necessary for something to be called a pain.

I reject the ordinary view. I think typical presentations confuse aspects of pain proper with states evoked by pain. I think imperativism can do with one simple component what the ordinary view does with two complex ones. But most of all, I think the ordinary view is wrong about what pains are *for*. The biological role of pain is a homeostatic one. Like hunger or thirst, pain is there to get you to act in ways that bring your body back into balance. Returning to balance only requires taking the right sort of actions. Your body doesn't need to tell you *why*—that information would only get in the way.

I will defend my picture at length in the chapters to come. As a brief illustration, however, consider my ankle. It aches. That aching keeps me from walking on it. I'm not sure why it aches. Maybe something is still torn down there, maybe it's just weak and needs special care. I don't know. Nor does it matter—whatever the reason, I should not walk on my ankle. If I can manage that, whatever is wrong will sort itself out. If I do walk on my ankle, it could make things worse, maybe permanently. Rather than leave it up to me, my body just motivates me not to walk around if I don't have to. Then I will heal. That is the function of pain.

This is why I will focus on pains of recovery. Pains that accompany recovery are the most prevalent, and arguably the most important, of the pains we feel. Philosophers often focus on pinpricks and other brief, acute pains. This has always struck me as slightly odd. The pain of a single sprained ankle will last longer than the total duration of all the pinpricks I will ever feel. Ignoring a pinprick has few consequences, and the threat is often gone before I notice. The pains of recovery, in contrast, are clearly adaptive and shape our actions for an extended time. It is obvious why they're there and obvious what they do. Focus on those, and imperativism already becomes more attractive.

1.2 Reflections on Imperatives

Imperativism sets up a parallel between pains and imperatives in ordinary language. I mean this quite literally: pains have the same kind of content, and play the same kind of role, as imperatives do.

A potential misunderstanding is worth cutting off at the outset. When I express the content of pain imperatives in English, I assume that the imperatives are instantiated in some form of mentalese, not natural language. Differentiating such imperatives from (say) *thoughts* with the same content is an interesting and complex project. I note that it's also one that *any* intentionalist theory faces, and so it has been rather widely discussed.[3] I will say more about this in chapter 3; for now, as long as you think that sensory content can be

3 For a good review of possible responses, see Egan (2006). I'm most fond of solutions on which the imperative content of pain is sufficiently complex to distinguish it from other sorts of mental content. That's not implausible: the imperatives of pain proscribe against a wide range of complex

suitably distinguished from other sorts of mental content, imperativism raises no new or distinct problems.

Philosophers do not work with imperatives as much as they used to. That's partly due to changes in the profession. Thinking about imperatives has a rich philosophical history, but the rise of deontic logics alongside the troubles with developing imperative logics have made imperatives less attractive as a technical device.[4]

Yet imperatives play an important role in many areas of life. Parents regularly command children. Parenting techniques have moved away from an emphasis on obedience, but sometimes there is no substitute for a well-timed "Stop!" Formalized structures of commands play an important role in the military, of course, and similar hierarchical institutions. Informal commands can also motivate us, as when the trainer exhorts us to do one more lift. Commands don't require hierarchy, however. Two people can both be willing to accept commands from the other, and that willingness is crucial for regulating various kinds of joint projects. If you and I are moving a couch in a cramped house, then our exchange might consist entirely of commands to one another: "Higher!" "Drop your end!" "Now turn!" Similarly so for giving directions, or baking together, or all manner of mutually enjoyable activities.

Start with moving a couch: this strips away many of the complexities that arise when there are hierarchical relationships. Why are commands so natural? Conversely, why is it frustrating to move a couch with someone who chatters away about how they would like to move a couch, or what one should do in this situation, or what they'd prefer as far as couch-moving is concerned?

The answer, I suggest, is that imperatives are directly and nondeliberatively action-guiding. When I say, "Lift your end!" and all goes well, you just lift it. You don't need to deliberate about what would be best or right or good or what I might want. If you're committed to accepting my command, my command is enough to get you to act. We save time, and we act smoothly. Coordinated action in cases where there's good mutual information works best when there's as little deliberation as possible. Indeed, Lewis suggests that the distinction

ways of bodily action and so differ from thought imperatives just in virtue of their additional content. For a solution crafted especially for imperativism, see Martínez (2010).

4 Although not entirely. See, for example, Vranas (2008, 2010) and Parsons (2012a) for recent work in philosophy and Portner (2007) in linguistics. Hamblin (1987) is an excellent (although opinionated) introduction to the history of philosophical treatments of imperatives.

between indicatives and imperatives is that the former require deliberation on the part of the receiver whereas the latter don't.[5]

It is precisely this direct link between imperatives and actions that makes imperatives so efficient and useful. This is why they are common in hierarchical situations, where the deliberation of the agent is presumed to be precluded by that of his superiors. The extreme case of this, suggests Raz (1986), are the commands issued by societies in the form of laws. Laws are commands meant to encapsulate the deliberative wisdom of the state—and thereby preclude further deliberation on the part of citizens.

Of course, commands don't always have their desired effect. You might reject my authority. Commands can be as confusing and ambiguous as any other form of communication. You might treat my command as merely an expression of my preference, a threat, or a warning. From there, you might deliberate about what you should do.[6]

I take it, however, that these are all *failures* of command: if I balk, if I stop to deliberate, I have ceased to treat the command *as a command*. Contrast this with cases where an imperative fails because I can't carry out what you've ordered, circumstances are unforgiving, or I'm in the grip of some other, more pressing command. There the connection between command and action is preserved but blocked. There you *have* successfully commanded, and, had other things been easier, I would have done as you said. In such cases, I am motivated, but that motivation is overridden.

In ordinary cases, then, the function of imperatives is to provoke action rather than deliberation. That is why commands are not truth-apt: what is commanded is to be *made* true, and the interesting division is between commands that have been satisfied and those that haven't. When I flesh out the semantics of commands in chapter 5, I'll thus focus on *satisfaction conditions* rather than truth conditions.

1.3 Two Background Commitments

I am a pure imperativist, and I am an intentionalist about phenomenal character. Both commitments make imperativism harder to defend and so constrain and shape my account.

5 See Lewis (1969, 42ff) and discussion in Huttegger (2007).

6 On general considerations against the *reduction* of commands to threats or other legitimacy-carrying representations, see Parsons (2012a).

First, I defend *pure* imperativism. The pure imperativist claims that pains *only* command. They have no (psychologically relevant) content beyond their imperative content. Contrast this with *hybrid* imperativism, which claims that pains have additional content over and above what they command. Hybrid imperativism could take pains to be simple conjunctions of commands and representations, or they could appeal to more complex contents that both motivate and inform.[7] As I write, imperativists are rare, and pure imperativists rarer still. Hall (2008) and Martínez (2010) both defend imperativism. Hall is explicitly a hybrid imperativist of the first sort about pain, whereas Martínez suggests the second route.

Second, I will assume that some form of *intentionalism* about phenomenal character is true. Intentionalism is, minimally, the claim that the phenomenal character of an experience supervenes on its intentional content (Byrne, 2001). This means that there can be no change in phenomenal character without a change in intentional content.

Intentionalism is often run together with *representationalism*, the claim that phenomenal content supervenes on how a state represents or describes the world. I accept that representationalism is true for some kinds of mental states. Sensations in visual modalities, say, have *indicative* or *representational* content. These modalities depict the world as being a certain way, and sensations in these modalities are veridical or nonveridical just in case the world is that way.

Representing the world is only one way of having content. Just as sentences can be indicative, imperative, or subjunctive, so mental states might come bearing a variety of different types of content. Imperativism simply claims that the subvening content in the case of pain is a *command* rather than a proposition. Here the contrast with vision becomes obvious. While vision says that the world *is* a certain way, pain tells you to *make* it a certain way. Although both have content, they have a different direction of fit. Vision has the sort of content that can be true or false. Pains have the sort of content that you can satisfy or fail to satisfy. Hence, the difference.

7 For example, along the lines of Millikan's pushmi-pullyu representations or Clark's action-oriented representations. Once upon a time I was more keen on these complex contents and defended them in early drafts of my (2007). Tim Bayne (2010a), although not explicitly concerned with pain, suggests something like a complex hybrid account in passing. Manolo Martínez has also indicated to me in conversation that he's attracted by this line, given the utility of Millikan's account in other areas.

Intentionalism is controversial. I am not interested in defending intentionalism (except perhaps indirectly by neutralizing a problem case). I think intentionalism is the best theory going for how to naturalize phenomenal character—I don't see how *else* you'd manage. As a naturalist, I suppose I'll stick with the most promising option. In any case, my adoption of intentionalism should not be objectionable given my aims. For one, phenomenal character is mysterious. Rather than remain mute and passive in the face of that mystery, we might as well soldier on and describe what we can describe in the best way we can. Content is a good place to start, and we can say a lot of interesting things about the content of pain without solving that deeper mystery.

What's more, the combination of intentionalism and pure imperativism makes my job much harder than it would be otherwise. The two together restrict my explanatory resources down to a single source: the imperatives that constitute pain. I will have to use that one resource to do a lot of things.

I have mentioned, for example, that pains are distinguished from ordinary sensations by their imperative content. But this is not the only distinction that needs to be drawn. Pains are not the only imperative sensations. I think sensations like hunger, thirst, and itches also admit of an imperativist treatment. Each are sensations that motivate rather than inform. They are sensations that play a crucial role in keeping us alive and intact, and they do so precisely by commanding us to act in appropriate ways. Although I will have less to say about these imperative sensations, I must still tell a story about what distinguishes them from pains.

Pains also differ from one another. So while every pain has some character in virtue of which it counts as a pain, different pains are distinguished by their more specific features. A good theory should account for that variation as well. There are three important phenomenological axes along which pains vary. First, *location*. The pain in my ankle is distinguished from the pain in my wrist by its felt location. Location has a clear connection to motivation: the pain in my ankle motivates me to protect my ankle, not anything else. Protecting my ankle might have implications for what I do with other parts of my body (to hobble, I also have to move many muscles in unusual ways). These other actions are secondary to, and in service of, protecting my ankle. Second, *intensity*. The pain in my ankle is strong, whereas the pain in my wrist is mild. The intensity of pains has a clear connection to motivation as well: strong pains motivate me to take action more vigorously than weak ones do. Third, *quality*. Pains can be sharp, burning, wrenching, or aching. The McGill pain questionnaire, a standard instrument for rating pain, contains seventy-seven different descriptors of this sort across a variety of categories (Melzack, 1983). I will argue that pain quality depends on *how* you should protect something: what differentiates an ache from a stab is the fine-grained details of the activity that's commanded. As a pure imperativist, I must argue that each of these differences between pains can ultimately be located in differences in the content of the command expressed by pains.

The above is just a sketch; later chapters will flesh out the details. For now, note that many of these problems are much *easier* if you are willing to abandon either intentionalism or pure imperativism. The pure imperativist has to grapple with problems that additional representational content solves trivially. For example, felt location takes a bit of work for a pure imperativist because commands aren't located. Adding representational content would make things easy—you can just locate the pain where whatever's represented happens to be. Similarly, accounting for pain's motivational power is easier if you relax the intentionalism and let yourself appeal to broader functional roles.[8]

Nevertheless, I think the combination of intentionalism and pure imperativism is worth defending. For one, I think it can be made to work, and that's interesting in its own right. Even if you end up thinking that pure imperativism fails, hopefully I can convince you that it has fewer shortcomings than you might have thought. Any additional content would thus play a smaller role. By trying to make pure imperativism work, and doing so in a way that is consistent with intentionalism, I intend to show how philosophically powerful an appeal to imperative content can be. That's why restricted accounts can be of philosophical interest.

1.4 The Plan of the Book

The goal of this book will be to defend pure imperativism. Because it is a relatively new account, I take my primary charge simply to be to show that imperativism is a *coherent* picture. That goal will take a fair bit of work: various objections have already been raised to it, and I've come to think that imperativism works best when presented as a whole package of views rather than piecemeal. Presenting that package, however, also requires motivating imperativism and giving some arguments in its favor.

8 Perhaps by adopting some form of quasi-intentionalism. See Ganson and Bronner (2013) for that sort of argument.

Most work on pain assumes that pain serves as a *symptom* of damage. I know that my ankle is sprained because of the pain, and that pain (somehow) gets me to act. In that sense, I am on an epistemological par with my doctor: she also knows that my ankle is sprained because I am in pain. My access to this symptom is a bit more direct than hers: unlike with the swelling, she can't just look and see. As far as what my pain signifies, however, and what should be done about it, the doctor and I are on a par.

I think that's absurd. If pain worked like that, then we wouldn't last long. The doctor can ignore my pains, can encourage me to walk despite the pain, and so on. The link between symptom and proposed action could be tenuous and arbitrary. My pain stands in a more intimate relationship to me and my ankle. It doesn't simply provide one piece of a riddle that I then solve on my own. Rather, my pain motivates me to do something about the problem. That's the biological advantage of feeling pain (and imperative sensations more generally): pain forces us to do the right thing. So here's the story in a phrase: in ordinary cases, pain is a not a *symptom* but *part of the cure*.

To motivate pure imperativism, then, I'll start by thinking about the biological role of pain. On this model, pains belong to a class I will call the *homeostatic sensations*: sensations that motivate action to preserve some bodily parameters within acceptable limits. Others of this class include thirst, hunger, itch, and dyspnea (the felt need to breathe). Each of these sensations motivates rather than informs. I will present a model of imperative content which shows that such intrinsically motivating sensations are best understood as imperatives (chapter 2). Then I will argue that pain belongs to this class (chapter 3).

Doing so requires dealing with a fundamental ambiguity in the term "pain" as it is used in ordinary English. R.M. Hare (1964) suggests that there are three distinct uses of the word "pain."[9] First, "pain" is used (as I will use it) to refer to the bare sensation that is common to pinpricks, postural adjustments, sprained ankles, and the like. This is the sense that I care about and the sense in which pain is a homeostatic sensation. This first sense carries no implication that the pain actually *hurts*, however. Indeed, I'll argue that there are common, nonpathological cases of pains that don't hurt or feel bad (just as there are hungers that don't hurt or feel bad).

Second, "pain" is used to refer to that sensation *plus* the accompanying felt badness. Third, some words pick out sensations that *hurt* but don't involve any pain in the first sense: this is what we mean when we talk about the pain of

9 See pages 94–5. I depart slightly from Hare's specific formulations.

heartbreak, the anguish of grief, and so on. As a paradigmatic case of the third use, consider Keegan's (2011) description of the English troops at Agincourt:

> The English had sought by every means to avoid battle throughout their long march from Harfleur and, though accepting it on 25 October as a necessary alternative to capitulation and perhaps lifelong captivity, were finally driven to attack by the pains of hunger and cold. (115)

Keegan's invocation of "the pains of hunger and cold" picks out not physical pains but rather other states that, when intense, are painful. Indeed, English is even messier than this. To Hare's divisions, we might also add a fourth use of "pain" and "hurt" to pick out the *cause* of a pain, either externally ("These tight shoes pain my feet," "Being bitten hurts") or self-generated ("It hurts when I move my arm").

In what follows, I'll reserve "pain" for the first of these uses. Chapter 4 will sort out the distinction between pain and suffering. The first use may actually be the least common one in ordinary discourse. Hare (1964) wryly remarks that it "would be so used more often if the occasions on which we have the sensation without disliking it were not so uncommon" (94). Nevertheless, I take it that once the distinction is made, this usage should be unobjectionable. Which is not to say that the *distinction* is unobjectionable: some readers will think it obvious, and others will remain unconvinced. If you're in the latter category, I offer the distinction provisionally, with the utility of the system to follow as a further argument for adopting it.

Having made the distinction, I'll then argue that pains are exhausted by their imperative content. This will require spelling out the general content of pains (chapter 5) and explaining how pains motivate. Briefly, pains motivate because commands motivate. That is why we give commands, that is how we treat commands, and that's what determines the success conditions for giving and accepting commands. A story about commands should thus shed light on the sense in which pains are intrinsically motivating. Chapter 6 fleshes out that story, showing how pains motivate by giving us reasons for action. I suggest that pains motivate because we accept our bodies as authorities and so as a source of legitimate commands. Pains also differ from one another in a variety of ways. Chapters 7 and 8 show how this variation can be accounted for by the intrinsic content of pains. The virtues of the account may not be obvious. Chapter 9 will defend it against a variety of objections, while chapter 10 will show the advantages of imperativism over accounts that assimilate pains to other intrinsically motivational states.

Imperativism claims that pains are intrinsically motivating. Chapters 11 and 12 will be devoted to dispatching putative counterexamples to this claim. I will show how the pure imperativist can treat otherwise puzzling cases in which people seem to feel pain but are entirely unmoved by it. Spelling out that story will also allow me to make precise the sense in which pains are intrinsically motivating.

An important part of the motivational story will be to distinguish the senses in which pains *don't* motivate intrinsically. Following the distinction made in chapter 4, I'll argue that motivation due to *hurt* or *suffering*—that is, the motivation to remove the pain itself—should not be considered as part of the intrinsic motivational force of pain. I'll further motivate that distinction by showing that it permits a recursive structure that allows for explanation of complex phenomena like the masochistic pleasures (chapter 13). Finally, I'll conclude with a discussion of suffering. In chapter 14, I'll show how imperativism sheds light on the typicality of suffering in response to pain. That is, while pains don't necessarily hurt, they normally do: imperativism can say something about why and give a more sophisticated story than is usually available to those who think that suffering is only contingently related to pain. Thus, while pains promote a particular biological end, they do so in a way that causes more sophisticated creatures like us to suffer.

2 Homeostatic Sensations and Imperative Content

2.1 Behavioral Homeostasis

Life exists on a thin margin. Survival requires many physiological variables to be kept stable within a narrow range. As neither we nor the environment are static, our bodies must constantly adjust to changing conditions. *Homeostasis* is the collective name for the processes that maintain this balance (Cannon, 1932).

Homeostatic processes may be provisionally divided into two classes. First, there are homeostatic processes that don't require us to take action. Our bodies regulate blood pressure, plasma pH, and renal creatinine levels. These processes can go on without our help or awareness. Second, there are homeostatic processes that require us to do something to bring our bodies back in line. If I don't have enough oxygen, then I must take a breath. If I don't have enough water, then I need to drink. In each case, there is a threat to bodily integrity, and eliminating the threat requires me to take some action.

Call this second class of processes *behavioral homeostasis*. Behavioral homeostasis is not sharply distinguished from the nonbehavioral kind. Many bodily variables are controlled by a mix of both. Core body temperature is regulated by both involuntary shivering and moving to a warmer locale. Water levels are regulated by both involuntary control of sweating and voluntarily drinking more water. Nonbehavioral processes can often control certain variables within relatively circumscribed ranges, whereas behavioral homeostasis must take over to deal with more extreme or long-term deviations. Behavior can also help when deviations are merely anticipated. Careful diabetics maintain their blood sugar by a series of deliberative acts, and most of us bring a coat when we know it is cold outside.[1]

Behavioral homeostasis can thus take many forms. Some behavioral homeostatic processes, however—and the ones that I'll be concerned with—are mediated by particular sensations. Hunger, thirst, and dyspnea are familiar examples. The feeling that your body is too hot or cold (as distinct from located tactile sensations of temperature) also promotes crucial regulation of core temperature (Craig, 2002). Each of these sensations motivates you to take some action. Taking that action promotes a return to an acceptable mean (all things being equal). So hunger promotes eating, thirst promotes drinking, feeling cold

1 Compare Minsky's (1988) claim that "To serve as useful 'warning signs,' feelings like pain and hunger must be engineered not simply to indicate dangerous conditions, but to anticipate them and warn us before too much damage is done" (286).

promotes finding shelter, and so on. Call sensations of this sort *homeostatic sensations*.[2]

Also plausibly treated as homeostatic are general states such as fatigue and nausea. These are felt as states of the whole body and prescribe actions (resting, staying still) done by the whole body. Some homeostatic sensations have a more specific felt location. Itches are the most obvious ones: they get you to scratch a certain area (Hall, 2008). Certain kinds of muscle cramps might also be treated as located homeostatic sensations: they keep you from contracting a particular overworked muscle group so that it can rest.

Not all homeostatic sensations have a name. The need to urinate and defecate are associated with distinct feelings, but they don't have particular English names. Often this is a quirk of English—the Germans have *harndrang* for the desire to void one's bladder.[3] Some of the feelings above come in specific varieties: there is a hunger for salt and for protein (Denton, 2006), although these also aren't common enough to warrant specific English names. While I focus on homeostatic sensations associated with health, not all homeostatic sensations need do so. The primary function of homeostatic sensations is to maintain a parameter within an acceptable range, and parameters needn't be health-promoting. The craving for a cigarette promotes smoking and so a stable plasma concentration of nicotine. Yet this is hardly healthy. Nor are all homeostatic sensations innate—a craving for cigarettes is again an obvious example.

Caveats aside, most homeostatic sensations play a crucial role in our survival. We eat and drink primarily *because* we get hungry and thirsty. One can learn to eat regularly without the sensation of hunger. But that takes a fair bit of planning and rational forethought and isn't terribly efficient. Eating when you're hungry, by contrast, can be done even by simple animals. Homeostatic sensations, by motivating situationally appropriate actions, are thus an efficient method for ensuring behavioral homeostasis.

2 Here I owe a debt to Derek Denton (2006) and A.D. Craig (2003a), who argue for pain's role as, respectively, a "primordial" or "homeostatic" emotion. I'll explain in chapter 10 why I don't like treating pains as *emotions*; aside from that, however, I think that Craig and Denton are basically correct about pain's role and its relationship to the other homeostatic sensations.

3 Thanks to Julia Staffel for the example.

2.2 Features of Homeostatic Sensations

Homeostatic sensations share several important features. First, homeostatic sensations are strongly and directly *motivating* under ordinary circumstances. Hunger motivates you to eat. It does so directly and immediately. To be motivated to eat, you don't need to engage in any deliberation about what hunger means, what it's telling you to do, or what the proper response to hunger might be. You get hungry. In virtue of that sensation, you gain motivation to eat. The strength of your motivation lies more or less in proportion to the intensity of your hunger.

That is not to say that you'll drop everything as soon as you're hungry, of course. Except in extreme cases, hunger is only one motivational element among many. Usually you must deliberate about whether to act on a state such as hunger. Perhaps you are on a diet. Perhaps you have only terrible food available. Perhaps you're waiting for surgery and have been forbidden to eat. In these cases, you might well decide that you have a stronger reason *not* to eat, and that stronger reason might win the motivational contest. Note, however, that this stronger reason does not eliminate the motivation that comes from hunger: your decision to refrain from eating comes *despite* your hunger rather than eliminating it.

That's for good reason. Homeostatic demands can't be put off indefinitely. Hunger signifies a state that needs to be resolved sooner or later, on pain of death. Few demands will thus outrank severe hunger, for the obvious reason that most other things you might want require being alive to get. Whatever your other desires, then, the homeostatic sensations must remain nonoptional parts of the motivational *milieu*. Severe hunger is typically only outranked by something that will kill you even more quickly (e.g., some other, more pressing homeostatic demand).

Second, homeostatic sensations motivate *types* of action rather than specific particular actions: eating, drinking, smoking, or scratching. Some homeostatic sensations promote a relatively flexible range of activities: you can warm yourself by seeking shelter, starting a fire, or putting on a jacket. Others are less flexible: thirst requires you to drink some liquid, although which one is up for grabs. Even sensations that demand only a limited range of actions still permit flexibility in how those actions are performed. Dyspnea might require inhaling, but that inhale can be done quickly or slowly.

These first two features also distinguish behavioral homeostasis from classical reflex behavior. Classical reflexes promote a single contextually fixed behavior in a relatively inflexible way. Homeostatic sensations, by contrast, give flexibility to an organism's behavioral repertoire. By loosening the connection between homeostatic imbalance and restorative action, they allow for delays that might be more adaptive. If a lion is prowling, then it's better to put off a trip to the river. Being able to hold your breath lets you swim. Being able to hold your bladder lets you swim in public pools. Decoupling homeostatic processes that can be satisfied by any one of several types of actions also permits adaptive planning: if I know the lion is at the river, then I can take a longer walk to the watering hole. Both would be roughly as good as far as my thirst goes.

Nevertheless, a close relationship exists between behavioral homeostasis and more reflex-like activity. As noted previously, homeostatic sensations often work together with autonomic mechanisms. The return to balance often involves a mix of the two. A.D. Craig (2003a) notes,

> Homeothermic mammals additionally regulate body temperature by modulating autonomic (cardiorespiratory) activity, which differentially controls blood flow to the thermal core and shell (the skin). Yet thermoregulation in mammals, including humans, necessarily still includes behavioral mechanisms; we not only use clothing, build fires and migrate to temperate climes, but also are motivated to respond behaviorally to immediate changes in the temperature of the core or the skin. (303)

So the best behavioral response is shaped by both internal conditions and the state of the external world.

Further, some homeostatic sensations are associated with top-down control of otherwise reflexive activity. The felt need to urinate, for example, is a product of toilet training—only when we learn to hold our bladder does the felt need to urinate really become pressing. Most "reflexes" are under the control of top-down commands from the cortex and are at least partially inhibitable by them. The need to urinate, for example, develops over time and is associated, as are many homeostatic sensations, with modulation via the periacqueductal gray (Fowler et al., 2008).

The linkage between motivation and action-types is partly constitutive of the homeostatic sensations. The whole point of homeostatic sensations is to get you to *do* something. In ordinary circumstances, doing that thing will remove the threat that caused the homeostatic sensation in the first place. Different threats require different responses. Hence, different sensations are associated with different action-types.

As Richard Hall (2008) notes:

> The connection between itches and scratching is very intimate. One is tempted to say it is a conceptual connection. . . . And if you look up 'itch' in your dictionary, the entry will almost certainly make prominent reference to scratching. (525)

Hall claims, and I agree, that this tight connection is something that a good theory of itches—or of any homeostatic sensation—should explain. Similarly, homeostatic sensations are linked to the particular kind of action they promote. What distinguishes itch from thirst is that itch motivates scratching, whereas thirst motivates drinking. Just as it's hard to imagine an itch without a push toward scratching, it's also hard to imagine an itch that makes you want to *drink* rather than scratch.

The first two features of homeostatic sensations go some way toward explaining why behavioral homeostasis is mediated by sensations in the first place. Nonbehavioral homeostatic processes go on in the absence of awareness. That's because there is no voluntary action that you can take which will reliably regulate renal creatinine levels (say). When there's nothing that an organism can do about a state, it makes little sense to have a sensation that is associated with it.

Homeostatic sensations thus represent a halfway house between mere reflexes and full agential desires: although they motivate action, they do so in a way that allows for deliberation and other sorts of top-down control. Conversely, doing the right thing—eating when you're hungry—is entirely sufficient to deal with the problem that gave rise to a particular homeostatic sensation, at least in ordinary circumstances. The deliberation needed is at best about *when* and *how* to satisfy the homeostatic demand. Homeostatic sensations give just the right amount of flexibility: flexibility as to *when* the homeostatic demand is satisfied and *how* (out of a relatively small range of actions) you satisfy it, but not as to *whether* you act or *what* (in a broader sense) you do to satisfy it.

A third feature of homeostatic sensations is that they tend to be relatively *uninformative* about their causes. Having a homeostatic sensation doesn't tell us much about the state that causes it. Even if we feel hungry every day, we don't thereby get more insight into the state of our body that causes hunger. One might *infer* a cause in specific cases: I itch, and I know I touched poison ivy, and so I infer that the poison ivy caused my itchiness. But this does not require thinking that the itch informs. I infer something about the cause of my itch from *the fact that I itch* and some associated beliefs about that fact, rather than from anything conveyed by the itch. It is textbook physiology, not introspection, that tells us about the causes of homeostatic sensations.

When we do learn the relevant physiological facts, we are often surprised by what we find. The sensation of dyspnea, for example, is caused not by the need for oxygen but by the buildup of carbon dioxide. The two go together in ordinary circumstances but can be dissociated. Inhaling a mixture with more than the ordinary amount of carbon dioxide will produce an intense sensation of suffocation, even if there's still sufficient oxygen in the mix (Meduna, 1950). That's not obvious and certainly not revealed by introspection.

What we *are* aware of is the action that a homeostatic sensation motivates us to perform. I may not know much about the underlying physiology of hunger, even vaguely. I do know, however, that when I'm hungry, I need to eat. Indeed, if you press someone to specify the bodily state to which a homeostatic sensation corresponds, then you'll likely get a state defined as the need to take the relevant action-type. What causes hunger? Needing to eat. What causes thirst? Needing to drink. In picking out the relevant bodily state, we simply describe the actions that would restore homeostatic balance, rather than giving some positive characterization of the underlying state.

Our handle on the homeostatic sensations therefore comes through being aware of the actions that they motivate. That is an obvious disanalogy with sensory modalities such as vision or touch. There, people are quite happy to refer the causes of their sensations to *things in the world.* Blue sensations are caused by blue things, smells by nearby odorants, and so on. Of course, some science or philosophy might cause you to revise your conception of the typical causes of sensation (including whether there are stable typical causes at all). But visual sensations are typically described by talking about states of the world (Smart, 1959). Again, this is in contrast to homeostatic sensations, which are typically picked out by referencing certain *actions* that have or haven't been taken.

Why the difference? At least two considerations might be relevant. On the one hand, most homeostatic sensations bear a degenerate relationship to bodily states. Hunger can be caused by a drop in a variety of nutrient levels. Thirst can be caused by a drop in fluid levels or a rise in sodium levels. The same itch can be caused by any number of skin conditions and parasites—and, further, the set of conditions that might cause an itch seem to be open ended. Therefore, what's common to all hungers, all thirsts, or all itches probably can't be some natural property in the world. Each sensation has in common only that it motivates a particular type of action. Those actions thus provide the handle we have on the sensations and are why we describe the sensations in terms of the associated action.

How plausible you find this argument depends in part on how unified you think the nonhomeostatic sensations are. In the crude form above, I have assumed that sensations of blue are unified by their function of picking out blue things (Byrne and Hilbert, 2003). If you're less of a realist than I'm inclined to be, then there are still ways to make the point. Because it's made most strongly in the case of dismissing representationalism about pain, however, the full version will have to wait for chapter 3.

On the other hand, you might think that homeostatic sensations fail to inform because their role simply doesn't require it. Modalities such as vision have a relatively flexible linkage to behavior. The right action to perform when we see blue or hear a train whistle is completely contingent on the context. Complete contingency of response would be a disadvantage when it comes to homeostatic demands, however. There is only one type of action that's appropriate in response to hunger: eating. Whether you eat right at that moment, what you eat, and how you eat might depend on context. But the appropriate response is fixed by facts about your body.

Homeostatic sensations may be thought of as a bit like fire alarms. The primary purpose of a fire alarm is to get people to evacuate: that is why they're loud, annoying, and persistent. The inventors of fire alarms don't want people to think about whether a fire is present, gather further information, or hesitate in any way. Fire alarms go off not to convince you that leaving is a good idea but to get you to *leave*. Of course, you might wonder whether an alarm is a drill, you might (with difficulty) ignore it and stay in your office, and so on. But this, in an important sense, is not the intended response to a fire alarm. Fire alarms are there to get you moving, not thinking. So too, I suggest, with the homeostatic sensations.

2.3 Homeostatic Sensations as Imperatives

Homeostatic sensations are best understood as sensations with imperative content. Each expresses a command. Satisfying that command will, under ordinary circumstances, remove the condition that caused the homeostatic sensation in the first place. Thirst is caused by low fluid levels. Thirst commands you to drink. If you drink, then you'll raise your fluid levels. Your thirst will cease.

Two features of this definition are worth emphasizing. First, satisfying the command is sufficient to alleviate the underlying condition and restore the body to balance only under normal conditions. If you're in a cloud of carbon dioxide, then a deep breath won't help you or your feeling of suffocation. Conversely, if you sever your vagus nerve or take enough methamphetamine, then you won't feel hunger even when it would be appropriate to do so. Perhaps less obviously, homeostatic sensations can also mislead. Classically, the dropsical patient thirsts, although drinking will make his dropsy worse.[4] Eating with a perforated stomach will neither alleviate your hunger nor make you healthier. Homeostatic sensations are set up to promote the right behavior only under typical circumstances. Further, homeostatic sensations are made to do so quickly and reliably (and to do so with the resources available to simple animals, not just big-brained creatures). That constraint means that mistakes in odd circumstances are inevitable; nevertheless, satisfying the commands given by homeostatic sensations *in general* promote adaptive behavior.

Second, satisfying the command removes the homeostatic imperative only indirectly, by removing its underlying cause. Of course, we often say that we alleviated our thirst by drinking water, but this is ambiguous between a direct and indirect reading. I claim the correct reading is the indirect one: that is, drinking removes the condition, which removes the sensation. This indirectness could be challenged. Maura Tumulty (2009), in the course of objecting to imperativism about pain, says that, "In typical cases, it suffices, for a feeling of hunger or thirst to cease, that one comply with the imperative—that one eat or drink an adequate amount" (162). I think this equivocates. What removes the imperative is not satisfaction per se but rather the elimination of the underlying physiological cause of the imperative. The two coincide in ordinary cases: the satisfaction is a way of eliminating the underlying cause. However, satisfaction is neither necessary nor sufficient for the elimination of the positive imperative sensations. Insufficient, because satisfaction that does not eliminate the underlying cause does not eliminate the sensation. Pathological itches are not eliminated by scratching (Gawande, 2008). When Edward Adolph (1941) diverted water from the stomach of his esophageal fistulous dog, its drinking did not quench its thirst (1365–1373). Unnecessary, because removal of the

4 "Dropsy" is an old term for edema, which is caused by poor homeostatic regulation of the interstitial fluid. Because of the disregulation, drinking fluid will make the edema worse.

physiological condition will eliminate the imperative in the absence of satisfaction. When Adolph added water directly to the stomach of his dog, it ceased to thirst even without drinking.

Different homeostatic sensations can also be distinguished by the different action-types that they command. Variations *within* sensations can be similarly traced to variations in imperative content. Urgency of sensation is precisely analogous to the urgency of imperatives: a strong hunger expresses more urgent a command than a weak one does. Itches vary in location, and that variation is plausibly linked to the location that you're commanded to scratch. One can have specific hunger for different nutrients such as sugar, salt, or protein (Denton, 2006). Differences in these imperatives can be traced to more fine-grained differences in the imperatives. All hungers are commands to eat; hunger for protein is a command to eat some protein.

This suggests that the *phenomenology* of homeostatic sensations can be accounted for by appealing to their content, compatible with an intentionalist account. I think that's right. Further, I think that a form of the transparency thesis holds here as well: when you are aware of a homeostatic sensation, you are aware (only) of a command, issued by your body, ordering you to take action. I'll return to these theses in detail when I get to pain.

Of course, arguments from phenomenology are difficult. I therefore present a more direct argument from considerations about biological signalling systems. I will argue that the functional role of the homeostatic sensations is the dual of that of representational sensations and that this difference can be given a precise characterization using recent work on signaling systems.

2.4 Imperative Signals

I have spoken of imperatives and indicatives as if they formed two distinct and exhaustive classes of content.[5] But the categorization is not exhaustive. Some signals can equally well be characterized as demanding action or describing the world. Consider the shot of a starter pistol. It can equally well be described as *informing* the runners that the race has begun and *commanding* them to begin. The state of the world and the corresponding action it evokes are so tightly coupled that either description seems warranted. The starter signal is,

5 This section is partly based on joint work with Manolo Martínez. For a technical treatment, including game-theoretic analysis, see Martínez and Klein's "Choosing Between Indicatives and Imperatives" (draft).

in Lewis' (1969) terms, a *neutral signal*: both a signal-that and a signal-to (134ff).

Several authors have noted the ubiquity of neutral signals in biological contexts. Millikan's (1995) "pushmi-pullyu representations" and Andy Clark's (1995) "action-oriented representations" both fall into this category. In many cases, organisms can get by with neutral states that simultaneously represent the world and provoke some action. Neutral signals aren't *merely* causal linkages between input and effector (as, e.g., classical reflexes might be). Rather, I take it that neutral signals are content-bearing states. It is simply ambiguous whether that content describes the world or specifies an action.

Neutral signals are appropriate for simple organisms in what Kim Sterelny (2003) calls "informationally transparent" environments. Informationally transparent environments are ones where there is a relatively simple one-one link between reliable environmental cues and appropriate behavioral responses (20). Sterelny calls organisms that can depend on simple reliable signals *cue-driven* organisms. Sterelny observes that

> Cue-driven organisms will often struggle if ecologically relevant features of their environment—their functional world—map in complex, one to many ways onto the cues they can detect. Such organisms live in *informationally translucent* environments. If food, shelter, predators, mates, friend, and foe map in complex ways onto physical signals they can register, cue-driven organisms' behavior will often misfire. (21)

A tight coupling between state and action isn't flexible enough to handle all of the situations we find ourselves in, however. Sterelny argues that the evolutionary solution to informationally translucent environments is the development of *decoupled* representations. Decoupled representations are:

> internal states that track aspects of our world, but that do not have the function of controlling particular behaviors. Beliefs are representations that are relevant to many behaviors, but do not have the biological function of directing any specific behavior.... Though these decoupled states are not tied to any specific behavior, they are potentially relevant to many actions in a number of task domains. Decoupled representation makes an agent's actions sensitive to a greater variety of information sources. (29)

As Sterelny details at length, decoupled representations are functionally advantageous in complex, informationally translucent environments because they allow for flexible response to the same environmental property in different circumstances.

Decoupled representations arise, on Sterelny's account, when there is a certain kind of *degeneracy* in the linkage between world and action. As he notes, decoupled representations arise when there are *many* actions that might be appropriate to take to *one* sort of environmental state.

What I have been calling imperative sensations, I suggest, arise in the converse of this case. That is, there are situations where *one* type of action is the appropriate response to *many*, often diverse environmental circumstances. The homeostatic sensations are the most obvious case where this degeneracy arises. As Konrad Lorenz notes:

> there are many physiological mechanisms which are there for the sole reason of letting us know that something is wrong. We feel ill without knowing the reason. The very fact that we have only one term, "I feel ill," for a range of conditions based on different causes is extremely characteristic. (quoted in Minsky, (1988), 93)

But, of course, merely *signaling* that something is wrong would be of dubious utility. Most crucial bodily parameters can deviate from the optimum for a variety of reasons; they are brought back into balance with the very same kind of action. Homeostatic sensations, then, can be thought of as those that connect a heterogenous disjunction of problematic states with a single type of response that would be appropriate for any member of the disjunction.

The reason for having a *signal* for this sort of state, rather than a reflex-like connection between conditions and action, is that a signal allows for more flexible behavioral response. Imperatives come into a system with a variety of motivational states, including other imperatives. What is important about an imperative signal is that only one action is the appropriate response *to that signal*. Flexibility arises because the organism can decide to forgo the immediate satisfaction of some imperatives to better satisfy the entire set.

Imperatives are thus the functional dual of indicatives. Both arise in response to complementary kinds of degeneracy in the relationship between environment and appropriate action. Both allow for a flexible response in systems that have complex, context-sensitive goals. Neither needs to be taken as more basic than the other, and neither is more primitive.

Indeed, this simple distinction shows up even in basic game-theoretic accounts of signaling games. As I noted, the idea of a neutral signal is already present in Lewis. Any signal, argues Skyrms (2010), carries two types of information: information about the state that caused the signal, and information about the act that will be performed in response to the signal. Huttegger (2007), following Lewis, argues that the distinction can be drawn based on the relative amount of deliberation necessary by the sender and receiver:

> If the sender does not have to deliberate but the receiver must deliberate, then the signal is indicative. If the receiver must not deliberate but the sender has to deliberate, then the signal is imperative. If both descriptions are compatible with the signaling behavior of both players, then the signal is neutral. (415)

Indicatives and imperatives, argues Huttegger (2007), relate to the world in a different way. Imperatives have a direct relationship to action and only an indirect relationship to the state of the world; indicatives have an indirect relationship to action and a direct relationship to the state they signal. Zollman (2011) abstracts away from this account, showing that, more generally, indicatives are those signals where potential states of the world are preserved while information about acts is eliminated when we average over different instances of the signal; the reverse is true for imperatives.

I suggest that homeostatic sensations fall firmly on the imperative side. Homeostatic sensations have a direct relationship to actions: hunger leads to eating regardless of the state that caused the hunger. However, homeostatic sensations carry little information about their triggering states—for those states are diverse across a variety of situations, often depend on preceding stimulation, and so on. The imperativist might concede that homeostatic sensations carry some information about triggering states—surely there's some big, ugly disjunction of states that cause hunger, for example, and so trivially hunger carries the information that one of those states obtained. (That is true, note, for *any* imperative signal.) Nevertheless, the tight, direct connection to action and the loose, indirect connection to states of the world put pains firmly on the side of imperatives rather than in neutral ground or on the indicative side of things.

An appeal to imperative content is thus neither surprising nor philosophically unusual. Imperative and indicative content lie at opposite ends of the same spectrum. Indicative sensations let organisms deal with situations where the same state of the world should prompt different actions in different circumstances; imperative sensations let organisms deal with situations where the same action-type is demanded by many different states.

Homeostatic sensations clearly mediate the latter kind of case, and that's good reason to think that they are constituted by imperatives. Neither indicatives nor imperatives are conceptually primitive in this picture. However, imperative sensations may be more *widespread*, because behavioral homeostasis is required of any moderately complex organism.

I think the signaling perspective also solves a related problem about how signals with either representational or imperative content might evolve. As I have drawn the distinction, indicatives and imperatives are simply two poles of a spectrum, with stereotypical neutral signals as the intermediate case. Simple organisms with simple needs in transparent environments can make do with neutral signals. As organisms and their goals become more complex, however, both internal signalling strategies can evolve.

On the one hand, the need to track complex properties in translucent environments puts pressure on the development of some indicative-type signaling strategies, which ultimately results in decoupled representations. On the other hand, there will also be cases where the same threat to integrity, caused by a variety of potential sources, is handled by the same type of action. That will give rise to imperative-type signals.

Among these signals, I submit, we should include the pains.

3 Pain and Imperatives

3.1 Why Do We Feel Pain?

Philosophical theories of pain presuppose a theory about the biological role of pain. Imperativism is no exception. Reflection on the role of pain will provide the primary argument for imperativism.

Many theories presuppose that pain's function is to inform us about damage. They then go on to give an account of just how it is that pains can inform, what they inform about, and how being informative is consistent with all of the aspects of pain that don't seem informative. Whatever the philosophical virtues of such accounts, however, as biology, they fall flat. For one, many pains don't inform about damage as such: pains of potential damage occur before anything gets hurt, and pains of recuperation can outlast damage. These are adaptive pains. We would be worse off if we did not feel them. A good story about the biological role of pain should include the obviously adaptive ones.

For another, saying that pain informs about damage still doesn't say what pain is *for*. As an explanation, it leaves out the link between informing about damage and why that information would be useful. Perhaps the answer seems so obvious that elaboration isn't necessary. I don't think so. A stellar probe might register damage solely to know how close it has gotten to the sun, the ablation of its heat shield being a good measure of proximity. A power plant might register damage so as to inform workers, who will then deliberate about what to do. There are all sorts of possible functions that information about damage might play depending on the system and on the possibility for intervention. A power plant needs outside intervention and might need to be shut down. We have to help ourselves and do so in the context of an ongoing life, with independent goals and actions already in progress. In this complex state, the link between pain and damage information is far from obvious.

I'll begin, then, by clarifying what I take to be the biological function of pain. As with the phenomenology of pain, puzzles abound. Inability to feel pain is bad for you, and so pain must play some adaptive role. Yet some pains don't seem to be adaptive, and sometimes we don't feel pain when it seems like it would be good to do so. T.P. Nash (2005), arguing provocatively that pain has no function, points out that many pains couldn't possibly have an evolutionary function. The pain of appendicitis, he notes, "surely had no use prior to our ability to do an appendectomy" (148). Nor is pain necessary for either reflex-mediated withdrawal or learning (to which pain has a complex relation and can even serve as a "reward" under suitable conditions).

Conversely, pain often fails to appear precisely when you would expect it most. Many serious injuries *don't* cause pain, at least initially. Painless injury is best known from Henry K. Beecher's (1956) work on injured soldiers at the Anzio beachhead during World War II. Beecher noted that many of the soldiers he treated claimed to feel no pain from their injuries—neither initially nor for up to several hours afterward. Beecher was impressed by the contrast with his post-operative civilian patients, who routinely complained of pain from the moment they awoke from surgery.

Beecher hypothesized that painless injury was due to top-down cognitive effects. Anzio was part of a prolonged, costly engagement. Soldiers who were merely injured, Beecher reasoned, did not view their injuries as a bad thing. Injury meant evacuation from the front lines, which was good. That's why they did not feel pain. While Beecher's explanation is often reported, it cannot be correct. Painless injury is also common even among people who view their injuries as entirely negative. A study by Melzack, Wall, and Ty (1982), for example, confirmed that 37% of emergency room patients experienced an initial painless phase of the injury. This painless phase could last several hours. A surprising range of accidents turned out to have a painless phase—including car accidents, badly broken bones, and the accidental amputation of a finger. Even more patients reported initial pain, which then faded for a time. None them thought that their injury was a good thing.

As Wall (2000) also notes, there is also evidence of painless injury in wounded animals, who presumably don't have the same complex concerns that would permit Beecher's interpretation. Painless injury is not explained by lack of attention or distraction: Beecher's soldiers knew full well that they had been injured. Nor will it do to cite factors such as a "rush of adrenaline" or the like: that at best specifies a lower-level mechanism, whereas we are looking for the reason why that mechanism is the way it is. So, on the one hand, pain seems like it is important to survival. On the other hand, pain shows up when it is not useful and is absent during real threats.

I will argue that pain should ultimately be treated as a homeostatic sensation. Its role is to preserve the physical integrity of the body, particularly after injury. By the argument of chapter 2, therefore, we have reason to treat pains as imperatives. I'll begin by considering P.D. Wall's underappreciated theory about the role of pain and then expand it to make it fully general. That expansion treats pain as a homeostatic sensation. I will then bring the circle back around to imperatives.

3.2 Wall on the Role of Pain

Patrick D. Wall was a seminal figure in twentieth-century pain science. Along with Ronald Melzack, he developed the gate control theory of pain and was pivotal in developing transcutaneous electrical nerve stimulation (TENS) therapy for pain relief. Wall (2000) noted that the congenitally insensitive to pain do seem to live drastically shortened lives, and that their issues are not confined to rare conditions such as appendicitis. This testifies to the utility of pain. Yet Wall was also impressed by the frequency of painless injury. Struck by these two phenomena, he developed a novel theory of pain's role.[1]

Wall (1979b) denied that pain's role was to inform:

> Pain is taken not as a simple sensory experience signalling the existence of damaged tissue. The presence and intensity of pain is too poorly related to the degree of damage to be considered such a messenger. Pain is a poor protector against injury since it occurs far too late in the case of sudden injury or of very slow damage to provide a useful preventive measure. Instead it is proposed that pain signals the existence of a body state where recovery and recuperation should be initiated. This places the word pain in the same class as words such as thirst and hunger which signal not only a body state but also signal the impending onset of a form of behaviour. (304)

According to Wall, pain bears a degenerate relationship to the variety of states that cause it. Rather than inform, Wall suggests, the role of pain is to *limit movement*. Limitation is especially important after injury, a role shown most dramatically by those born insensitive to pain.

Wall noted that the congenitally insensitive to pain did not have the troubles that you might expect. That is, they did not, by and large, succumb to dramatic acute injuries. Like us, they have an interest in keeping their bodies intact. In the course of childhood, they mostly learn to avoid serious injury, just as we do. They need to take more care than we do—we can wait until we feel the heat of the stove, whereas they have to keep a close eye on the burner. Some also succumb to internal conditions for which severe pain is the primary diagnostic symptom—appendicitis, ectopic pregnancy, and cancer.[2] Put those aside. As Nash notes, these conditions are only treatable in the modern era, and so do not give much insight into the evolved function of pain.

1 What follows is based primarily on Wall (2000), also drawing on material from Wall (1979a, 1979b).

2 Nash (2005); see also Sternbach (1963) for a review.

Instead, Wall notes those who do survive childhood are undone by the combined weight of repeated, trivial, but unhealed injuries. In ordinary life, we suffer numerous small insults—sprains, scrapes, and minor muscle tears, more if you're clumsy. But even the most graceful among us can't avoid a bit of wear and tear. Pain, notes Wall, alters your movements to get you to protect injured body areas. This can be obvious in the case of dramatic injuries. But a subtle sort of modulation also happens with the smaller wear-and-tear stuff. The day after a long hike, your muscles are sore. You take it easy. That lets your muscles recover from their unusual exertion.

It is easy to overlook these minor aches and pains. The congenitally insensitive to pain provide a powerful example of the importance of both big and small pains. As Wall (2000) describes,

> With congenital analgesia, this recovery phase with guarding after minor injury does not occur. The consequence is that the surfaces of joints and ligaments never fully recover. Furthermore, the joint is in a particularly bad shape to counteract the next trivial injury.... The consequence of repetitive minor injuries with congenital analgesia is that joints, particularly ankles, knees, and wrists, become demolished. (51)

Infection eventually sets in, and death follows (Melzack and Wall, 1996, 18–19). The lesson? Pain's primary role—the one that can't be replaced—is to keep us from aggravating existing injuries. The other features of pain can be treated as derivative on this basic role. A system that keeps you from walking on a broken ankle will also (if possible) keep you from hurting it in the first place, because the same peripheral signals will be present in both cases, and the same sort of limitation will be adaptive. Conversely, it's hard to do without this protective function. Limiting motion keeps you from satisfying many of your other ends. Without a strong *motivating* sensation such as pain to stop me, I will naturally seek to satisfy my other needs at the expense of healing.

This perspective, argues Wall, also explains why painless injury occurs. Limiting movement can be maladaptive in life-threatening situations. Wall (1979a) suggests that painless injury is most common when the person needs to flee a dangerous situation or otherwise get help. Then, "as the necessity or possibility of destroying or avoiding the cause of injury subsides," pain begins (261). Wall's account is thus a *two-stage* theory of pain.[3] There is an optional

3 See Wall (1979a) for an early expression and Wall (2000) for a popular treatment. Wall (1979a, 1979b) sometimes spoke of three phases. The third phase is the onset of chronic pain, which is clinically important but clearly a contingent feature of pains. Hence, I think the two-phase version is more philosophically telling. Wall seems to have come to the same conclusion. The two-stage

first painless stage of injury, which permits flight, and a later mandatory painful stage. Beecher went astray, suggests Wall (2000), by comparing one group of patients immediately after injury with another much later in the course of pain. These populations are at two different phases of pain and should not be compared.

Note that Wall's position on painless injury can easily be misunderstood. It is not the case that one must *first* feel pain and suppress it at some later time. Nor must top-down suppression be mediated by personal-level states such as beliefs and desires. The pain system is complex and able to integrate information from a variety of sources (Melzack, 1990, 1999). Spinal reflexes are continuously modulated by top-down signals from the cortex. In fact, some spinal reflexes *never* manifest in ordinary life—even slightly—after otherwise appropriate releasing stimuli. This is most obvious in the case of the so-called "primitive reflexes" seen in infancy. The primitive reflexes are tonically suppressed by top-down signals in adults. They reappear after severe cortical damage, showing that the underlying spinal dispositions are constantly suppressed, rather than merely disappearing in adults (Plum and Posner, 2007; Schott and Rossor, 2003).

Hence, new stimuli do not provoke otherwise absent downward modulation—they merely change an ongoing modulatory process. Top-down modulation doesn't need to wait for signals to arrive from the periphery to be effective. That means that our response to pain can be complex and flexible—and absent, if that's most appropriate.

3.3 Pains as Homeostatic Sensations

Wall's two-stage theory is ingenious. Recuperation and repair are important effects of pain. I believe they are probably the most important functions: again, we spend far more of our lives feeling pains of recuperation than any other sort of pain. Recuperation is not the only function of pain, however, nor is it the only possible response to pain (although it is arguably the most common response). So I think Wall overplays his hand. Nevertheless, Wall's theory contains a crucial insight, which we can preserve in a more general form.

theory is distinct from the concept of "first" and "second" pain (Price and Dubner, 1977). First and second pain are pains with differing qualities, sometimes separated in time due to the differences of transmission time in Aδ and C-fibers. Wall's first stage, by contrast, is painless if it occurs. So both first and second pain must belong to Wall's second stage.

On Wall's story, pains motivate actions that will remove the threat that caused them. Pains limit motion by motivating you not to move. If you avoid moving, then you heal; the less you move, the faster you heal. All that can happen, as Wall notes, without pain providing any *information* about your body.

Pain, then, seems precisely analogous to the other homeostatic sensations. First, like the other homeostatic sensations, pain clearly plays a role in protecting the physical integrity of the body. The stable state toward which it tends is just one where the physical body is more or less intact and normally functioning (or, at least, as normally functioning as possible given the circumstances). The world pushes my body out of this state in various ways, and getting back into it requires altering my behavior. As a located homeostatic sensation, pain is most analogous to itch: both require taking action with respect to some particular part of the body to eliminate threats.

Second, pain achieves its homeostatic goal by promoting a particular action-type. Each pain, I suggest, motivates you to *protect* part of your body. Pains of potential damage and exertion motivate you to take actions that avoid damage in the first place, thereby protecting the threatened body part. Pains of actual damage and recuperation motivate you to protect an injured body part so that it might heal. Protection can take a variety of forms (although, following Wall, most will involve limiting movement in some way).

This accommodates Wall's key insight and fits nicely with his two-stage theory. Insofar as painless injury occurs, it's precisely when such protective actions would be maladaptive by prohibiting the necessary avoidance of dangerous situations. When safety should take priority over protection, injury may end up painless.

Third, acting on pain's motivation will actually achieve the desired result in ordinary cases. By protecting my ankle, I enable it to heal. So acting on the pain will remove the underlying cause of the pain. This is exactly parallel with other homeostatic sensations. Note that action removes the sensation only indirectly, as with other homeostatic sensations: return to ordinary functioning removes the pain, and pain behavior merely facilitates that return.

A fourth and final consideration involves the neural bases of pain. Treating pain as a homeostatic sensation would explain the significant neural overlap between pain systems and the neural systems involved in other homeostatic sensations.

A.D. Craig is the most prominent defender of the *neural* link between pain and general behavioral homeostasis.[4] Craig (2003a) notes that the neural pathways for pain are similar to those that carry thermoregulatory sensations, at both the spinal and cortical levels. There is considerable evidence that the insula plays a role in the awareness of homeostatic sensations (Craig, 2002; Singer et al., 2009). The insula also plays an important role in top-down autonomic control (Ibañez et al., 2010). Imaging studies consistently show a role for the insula in sensations of salt hunger, thirst, dyspnea, and the need to defecate (Craig, 2008; Denton, 2006). Conversely, damage to the insula can disrupt a variety of homeostatic sensations. The most drastic of these disruptions is the condition known as pain asymbolia, to which I return in chapter 11.

These common pathways suggest that pain is similar to homeostatic sensations. I don't want to place too much weight on these common pathways. The insula is also involved in numerous other functions (Singer et al., 2009), and I'll suggest later that insular damage does not affect sensations per se but rather the integration of those sensations into bodily care.

Nevertheless, we might make the following, more cautious argument. The fact of convergence is at least some evidence that pain plays the same *role* as the other homeostatic sensations. Why? Because information about pain must be routed in a similar way to similar structures. Conversely, there's good evidence that pain-relevant information does not appear to rely on the pathways that ordinary tactile information follows. Tactile information is processed in somatosensory cortices. Craig (2003b) notes that imaging studies do not reliably show somatosensory activation during pain (once appropriate regional boundaries are drawn); further, even large lesions to the somatosensory cortex do not reliably eliminate pain perception unless they also involve the insular cortex or the surrounding white matter. In summary, pains appear to do the same sorts of things as other homeostatic sensations and for the same reasons. Conversely, they *don't* appear to signal damage.

4 Craig considers these to be homeostatic *emotions*, a view I'll consider in chapter 10. I'll translate into my preferred terminology except in direct quotes.

3.4 Pains as Imperatives

3.4.1 The Argument

I have proposed that the biological role of pain is to get you to *protect a part of your body*. Pain can do this without informing because it is intrinsically an action-guiding signal. You feel pain so that you might take certain kinds of actions and avoid others. Each of these can be described as protecting a part of your body. I've suggested, of course, that this action guiding is done by the imperative content that constitutes pain: pain will end up being a *command to protect*. By obeying it, you protect your body, and pain thereby fulfills its role.

Given this, and the similarities between pain and sensations such as hunger, I suggest that we should treat pains as homeostatic sensations. In the previous chapter, I argued that homeostatic sensations have imperative contents. The main argument for imperativism about pain is simply a re-application of that argument. In particular, I claim that pains should be understood as imperatives, which express a command to protect a particular body part, in a particular way, with a particular intensity.

Linking this with the signaling argument of chapter 2 requires establishing two things. First, it requires that pains be associated with a unified type of action. I've argued that protective motions provide that unity, and the rest of the book will elaborate that claim.

Second, it requires pain to bear a degenerate relationship to its typical causes: that is, the causes of pain should be too heterogenous to count as unified for the purposes of signalling. I've claimed this in passing, but many theorists have denied that second claim. A representational view of pain is widely assumed and occasionally defended directly (most notably by Tye, 1995; see also Harman, 1990). There is, of course, considerable debate about whether pains *only* represent. Many assume that representations are insufficient, and so pains must be conjoined with an additional motivational state that forces us to act on that representation. There is also considerable debate about *what* pains represent (Tye, 2006), whether our ordinary concept of pain is as of something representational (Aydede, 2006a), and whether pains represent in some full-fledged sense or whether they merely indicate damage (Aydede and Güzeldere, 2002). That said, all these positions still assume that pains indicate something like tissue damage.

I deny this. As I've noted, pains are caused by an amazingly heterogenous set of states. That weighs against representationalism but does not yet settle the question. It may be that this variety of states has something in common. If so, then we might properly speak of pain as being caused by that common property and so carrying information about its presence. That's surely the case with ordinary representational sensations. A blue sensation can be caused by a bluebonnet, a bluebird, or a blueberry. Those are pretty different—except that they're each blue. Because they're all blue, everyone is happy to say that blue sensations carry information about a state of the world (viz.: its blueness).

Pain might be like that. Most theorists, I suspect, usually think pain *is* like that—that is, that there is some nontrivial property that pain-causes have in common and about which pain informs. I'm going to argue that there isn't such a property. I'll go over the usual suspects and explain why they don't provide the required unity.

A few ground rules. First, candidate unifying properties should be *nontrivial*. Very roughly, a non-trivial property should be the sort of thing that it would be worth carrying information about. (A trivial example would be that pains carry information about the presence of something that has caused a pain. True, but hardly illuminating!) Second, we're looking for something that *pains* have in common. It might be that individual pains carry more detailed information about what caused them. Maybe you think that stabbing pains carry the information that they were caused by something stabby. I remain neutral on that question: the issue at stake is whether pain *as such* carries information. Third, candidate properties have to be *non-disjunctive*. That's not just out of a general horror at the idea of disjunctive properties (although that horror is worth having). The name of the game is to show that there is some unitary property about which pains might inform. Given that, disjunctive candidates fail at the gate.

I'll proceed by exhaustion.

3.4.2 The Heterogeneity of Causes

3.4.2.1 Tissue Damage

Perhaps what unifies the causes of pain is that they are all instances of *tissue damage*. Falling off a roof, being bitten by a camel, inhaling ammonia, and turning your ankle don't have much in common at the gross physical level. But they're all ways of injuring your body. Philosophers are generally fine with such higher level properties, and information about tissue damage also seems like the sort of thing that would be evolutionarily useful to get.

To reiterate a point I've made several times, however, many perfectly adaptive pains don't involve actual tissue damage. Pains of potential damage, exertion, and recuperation are just as useful but not associated with damage per se. How to respond to this diversity? I can see two possibilities. First, one might go disjunctive: one can say that pains inform of tissue damage *or* potential damage *or* past damage *or*. . . . The goal was to avoid disjunctions, however; while this is better than an enormous disjunction of possible states, it's still messy enough to make this line a nonstarter.

Second, one might bite the bullet and say that pains *do* inform about tissue damage, in the sense that this is their primary function, and that other pains are strictly speaking malfunctions. This is extraordinarily counterintuitive, however. Pains of potential damage and recuperation are just as biologically important as pains stemming from tissue damage. I've argued that they are *more* so. Choosing tissue damage as *the* state about which pains inform thus seems arbitrary.

One might defend the claim that pains indicate tissue damage by appeal to ordinary talk about pains. When we wish to convey the quality of a pain, for example, we describe it in terms of some sort of damage—stabbing, burning, wrenching, and so on. The IASP definition of pain says that pain is an experience "associated with actual or potential tissue damage, or described in terms of such damage" (IASP Task Force on Taxonomy, 1994). Even if the imperativist offers an alternative treatment of pain quality, this line goes, the fact that we normally describe damage when we want to describe our pains suggests some important connection between pain and damage.

I think this is a red herring. For the fact that we talk about pain in terms of damage only shows that we use easily accessible public events when we want to talk about sensations. Because damage is a typical cause of pain, it's natural to use damage to teach the word "pain," just as we use publicly accessible objects to talk about our other sensations. Speaking of my tinnitus, I might say, "It's as if a mosquito is always buzzing about near my ear." That doesn't mean that my experience actually represents or indicates nearby mosquitos. Rather, I describe a state of the world that my interlocutor should be familiar with, invite him to think of what he would experience in that state, and then suggest that I am feeling something like that.[5]

5 This line of response is inspired by Smart's (1959) topic-neutral analysis of sensations. For a similar suggestion about pain, see Baier (1962).

Elaine Scarry (1985), remarking on the "as if" structure of pain, notes that, "Physical pain is not identical with (and often exists without) either agency or damage, but these things are referential; consequently we often call on them to convey the experience of the pain itself" (15). The subsequent explication of the IASP definition makes this clear, noting that, "Each individual learns the application of the word through experiences related to injury in early life." So regardless of whether pain indicates damage, the easiest way to *talk* about pain is by relating it to publicly available events such as injuries.

Indeed, I think it's a mistake to read the IASP definition as anything like a philosophical explication of pain. Rather, it's clearly intended as a *criterion* that lets doctors and researchers pick out pains when they see them.[6] Toward that more modest aim, I think it works just fine. The pain in my legs as I run may or may not have anything to do with damage; in most cases, I'm reasonably certain that my legs are fine. If I go to *describe* that pain, however, then I may well say that it's as if my muscles were burning. Reports like that are reliable indicators that I am in pain, regardless of whether the pain has anything to do with damage. So our talk of pain in terms of damage gives no evidence about what pains actually convey.

3.4.2.2 Disturbances: Setup

So "tissue damage" won't do. If there's any unified state, then it will have to be at a higher level of abstraction. Savvy philosophers use the term "disturbance" for that more abstract property. In most cases, I suspect, "disturbance" is not used with any further analysis: it's simply a placeholder for whatever unified state causes pain. That may be unobjectionable when the focus of theory-building is elsewhere. But it's clear that "disturbance" *needs* further analysis if it's to save representationalism.

Many things are disturbances of the body, in the ordinary sense of the word, but don't give rise to pains. Pressure, heat, and cold all disturb the surface of the skin from its ordinary state but aren't necessarily associated with pain. Most itches are caused by pathological disturbances of the skin, but itch is distinct

6 Similarly, the IASP claim that "pain is always subjective" needn't be read as a commitment to anything like the incorrigibility of mental reports or the denial that pain is also a physical phenomenon. Instead, it is clearly meant as a quite reasonable rule for practitioners: if someone says that they are feeling pain, then take their word for it. Hence, the formulation "should be accepted as pain," rather than "is pain" in the subsequent explication.

from pain. Conversely, light touch on an inflamed joint is painful but isn't a particularly striking disturbance (at least in the ordinary sense of the word).

"Disturbance" must therefore be treated as a technical term in need of explication. I've heard this cashed out in three different ways.

3.4.2.3 Disturbances: First Version

First, one might try to link "disturbances" to ordinary perceivers. That is, the cause of pain (considered as a class) is whatever causes ordinary perceivers to feel pain in ordinary circumstances. As a parallel, consider a dispositional analysis of colors: that is, colors are just dispositions to cause certain kinds of sensations in normal perceivers (Smart, 1959). Then it seems that colors carry information about things that cause color sensations. Saying that, however, isn't obviously trivial. It's a good bit of analysis. Why not say the same about pain?

I'm suspicious of appeals to "normal perceivers" when it comes to pain. The example of painless injury shows that the response to pain is individual and context-sensitive (as with inflammation). I'll outline further factors in this context-sensitivity below. Putting that aside, the logic of the color case is quite different from that being proposed for pain. Note that the normal-perceiver story about blue starts by *supposing* that there are things—colors—about which color sensations inform. Colors are mysterious and in need of analysis; the dispositional account provides one, leaving open how those dispositions might be further analyzed. Is it ultimately circular? I don't know. What's important is how different this is from the case of pain. For pain is problematic at the first step: whereas everyone is pretty confident that there are blue things (at least in ordinary, everyday ways of talking), the set of pain-causing things is heterogenous in ways that defy generalization. That's what started the hunt for a common property in the first place. So assuming that there's some common property is question-begging.

3.4.2.4 Disturbances: Second Version

A second possibility would be to link "disturbance" to something like the *homeostatic deficiency* that pain is meant to address. While the ways of going wide of the mark might be numerous, this line goes, they have in common that they are wide of some mark or other. On my account, pain is meant to bring one back to some stable state of bodily integrity. Why not just associate the relevant disturbance with the *negation* of that state?

This is also unpromising. For one, even if we take the homeostatic state that pain promotes to be a natural property, the negation of a natural property is rarely a legitimate property. Being a lorikeet is a perfectly natural property, whereas being *not* a lorikeet is a causal and scientific mess. The whole point of pains, I've argued, is to collect a diverse group of states, each of which demands the same sort of protective response.

For another, this move, if pressed, would erase the intuitive distinction between indicatives and imperatives. For *any* imperative to ϕ trivially corresponds to a state of needing-to-ϕ. So if this were a legitimate move, then the distinction would collapse. But the imperative/indicative distinction, in both the functional and contentful senses, is worth keeping.

3.4.2.5 Disturbances: Third Version

A third and final possibility focuses on the bodily mechanisms for pain perception. Pain is caused by signals coming from the periphery. So pains must carry some information about what's going on in the periphery, and hence must represent peripheral events. Call the peripheral systems dedicated to detecting painful stimuli the *nociceptive* system. We can identify disturbances with those peripheral events—and, in particular, with activity in the nociceptive system. Pains, on this view, would be caused by anything that activates pain receptors, or peripheral pain fibers, or something along those lines. Because our body has a system dedicated to picking up the states that lead to pain (the line goes), you can appeal to activity in that system to unify the causes of pain.

This argument could be run in two slightly different ways. On the one hand, you might think that the presence of a peripheral nociceptive system gives evidence that there's *some* unified property, because we only have sensory systems to pick up things that are properties in some reasonably robust sense. This single system would then pick out what's common to actual tissue damage, potential damage, and anything else you please. Perhaps there is some further interesting characterization of this set—but that the set forms a distinct kind is guaranteed by the existence of a detector system. On the other hand, you might simply identify "disturbance" with activity in the nociceptive system. This is similar to the "cause of pain" line above, but it's no longer trivial—for "nociceptive system" presumably picks out some complicated but objectively discoverable set of structures in the periphery, which are only contingently the cause of pain.

The success of this strategy hinges, however, on the existence of a nociceptive system with the appropriate properties. In particular, nociception must have a dedicated, specific set of pathways that causes pain reliably and only causes pain. Call this a *specificity theory* of the nociceptive system.[7]

Specificity theories were common from the mid-nineteenth through the mid-twentieth centuries. The specificity theory has its roots in Descartes but takes off in the mid-nineteenth century with Johannes Müller's "doctrine of specific nerve energies." Starting with the fact that nerves and nerve signals appear more or less identical throughout the body, Müller argued that sensations could not be differentiated by their intrinsic properties. Instead, sensations must be differentiated by the termination point of different nerve fibers in the brain, with different fibers carrying information to different centers. Müller's theory was elaborated at the end of the nineteenth century by Max von Frey, who gave us specificity theory in the more or less modern form. Von Frey argued that there are four types of cutaneous sensation: touch, heat, cold, and pain. Following Müller, he argued that each type of sensation was differentiated by the brain center to which sensory fibers projected. Each of the types was associated with specialized receptors in the skin. Pain was transduced by the most numerous of those receptors, the free nerve endings, and the fibers attached to hair follicles. After transduction, there was a straight shot from the periphery through the spine to the brain. Later refinements tried to associate particular pain qualities with subfeatures of this system, typically by looking for differences between pains caused by fast myelinated Aδ fibers and slow unmyelinated C-fibers.

Unfortunately, subsequent physiology has made this simple picture untenable. To be clear from the outset, the objection to specificity theory is not an objection to the physiological claim that the skin contains various specialized receptors and pathways. Rather, it is an objection to the purported link between physiological specialization and the creation of a specific sensation when and only when specialized structures are active. As Melzack and Wall (1996) put it:

> Consider the proposition that the skin contains "pain receptors." To say that a receptor responds only to intense, noxious stimulation of the skin is a physiological statement of fact.... To call a receptor a "pain receptor," however, is a psychological assumption: it implies a direct connection from the receptor to a brain centre where pain is felt, so that stimulation of the receptor must always elicit pain and only the sensation of pain. It

7 The presentation in the next few paragraphs follows closely that of Melzack and Wall (1996), especially chapters 8 and 9.

> further implies that the abstraction or selection of information concerning the stimulus occurs entirely at the receptor level and that this information is transmitted faithfully to the brain. The crux of the revolt against specificity, then, is against psychological specificity (155)

The simple relationship between peripheral firing and psychological effect was called into question by the development of gate control theory, which allowed a complex time-dependent relationship between peripheral firing and spinal output (Melzack and Wall, 1965). The complexity of peripheral response means that one should not expect differentiation between pain quality based on simple properties of peripheral receptors. Indeed, this appears to be the case. Wall and McMahon (1985), reviewing evidence from microneurography studies that stimulated peripheral fibers, concluded that there is no simple relationship between peripheral activity and pain quality. Most importantly for my argument, stimulating different fibers and receptors by a variety of means produced identical sensations (see also Melzack and Wall, 1996). While subsequent physiological work has led to a partial revival of the specificity theory (Perl, 2007), that has mostly focused on the presence of "pure" nociceptors and specialized pathways leading from the spinal cord. However, one must distinguish the proposition that the periphery contains specialized fibers from the *psychological* claim that these fibers have a neat correlation with pain quality. The revolt against specificity was against psychological specificity, and the evidence continues to weigh against specificity in this sense (Melzack and Wall, 1996).

In summary, "activity in the nociceptive system" turns out to bear a complicated and degenerate relationship to felt pain. It will not give the sought-for unity to "disturbance."

3.4.3 The Argument Recapped

Having canvassed what seem to be all of the plausible candidates for a disturbance theory of what pain represents, I doubt any theory will give that unity. Because disturbances were the most plausible thing that pains represented, we have reason to believe that pains represent no particular state. Conversely, pains *do* promote protective activity, which is just to say that, like the other homeostatic sensations, pains are better thought of as imperatives than indicatives.

The overall argument of this chapter proceeded in two steps. First, I argued that the phenomenological and functional similarities between pain and the other homeostatic sensations were a reason to treat pain as among the imperatives. Second, I argued that pains should be considered as imperative rather than indicative, given the framework sketched in the previous chapter. In particular, I argued that pains bear a degenerate relationship to states that cause them while bearing a relatively simple relationship to the protective activity they promote.

The two steps are separable: they are each meant to support one another, rather than to be either jointly necessary or jointly sufficient for the conclusion. Of the two steps, I'm less confident about the second than about the first. In particular, I do not think the argument of this section is a compelling argument against all forms of representationalism. It assumes, for example, that reflections on the relatively simple sort of signaling system inspired by Skyrms are sufficient to settle the issue. It leaves open the question of whether a more complex representational content might be sufficient to capture pains. Of course, the particular examples chosen may not convince everyone. I have assumed, for example, a relatively simple form of anthropocentric color realism inspired by Hilbert (1987), a view that is far from universally held. Further, it may be that color experiences in simpler or evolutionarily old animals might be more directly action-guiding, as championed by Matthen (2005).

The representationalist, in short, has many resources to respond. As I mentioned, my primary goal in this book is making imperativism plausible and attractive, rather than taking on all comers. Nevertheless, I think the anti-representationalist argument should have some bite to it. The sheer diversity of things that cause pain, combined with the relatively simple homeostatic demands in response to any particular pain, should give the representationalist pause. Representationalism is, I suggest, attractive precisely because it links sensations to content, and content is the sort of thing that seems like it might be naturalized. But that's true of imperativism as well, and there are cases where imperative content seems to be a better fit. The argument of this chapter is that pain is among those cases. The argument in the remainder of the book is, in essence, that imperative content both works and solves various puzzles; that's even more reason to find imperativism attractive.

3.5 Conclusion

Linking pains and homeostatic sensations concludes the direct argument for imperativism. Of course, plenty of details remain to be fleshed out. The remainder of the argument—at least in this book—is abductive (in the broad sense that applies to philosophical argumentation). Imperativism gives a satisfying, integrated account of pain phenomena. It solves many long-standing philosophical problems. That's why you should be an imperativist.

Imperativism's success as an abduction depends on marking off some phenomena as outside of the proper scope of a theory of pain—most notably, those that have to do with suffering rather than pain proper. Again, such negotiations are a familiar part of philosophical argumentation. With that in mind, I turn to a final bit of carving: to make the distinction between pain and the closely related but separate quality of *suffering*.

4 Pain and Suffering

"Pain don't hurt."
–Dalton, "Roadhouse"

4.1 Two Senses in Which Pains Motivate

Imperativism claims that motivation is an intrinsic feature of pains. In later chapters, I'll discuss how imperatives motivate, and thus how pains motivate. Direct motivation in virtue of imperative content is not the only way in which pains motivate, however. Pains do many things. For example, pain might make us anxious, and that anxiety might affect our behavior. That's not the main thing pain does, but it is an important secondary reaction *to* pain. So an imperativist should say something about these other ways in which pains motivate and distinguish them from the core motivational force of imperative content.

To begin, I distinguish two different types of motivational force.[1] *Primary* motivational force is that which stems from the intrinsic properties of pain. Imperativism claims that the primary motivational force of a pain is simply that which derives from its content: that is, the motivation to protect a certain body part, in a certain way, with a certain urgency.

Secondary motivation, by contrast, includes all motivational states that are properly extrinsic to pain. This includes any mental states that are *caused* by or *directed toward* the sensation of pain. Because secondary motivations are not intrinsic to pain, they can be entirely absent when we feel pain. Secondary motivation can also be contrary to primary motivation. I might be motivated to stretch a painful leg precisely because I know that stretching it will eventually remove the feeling of pain. This may be contrary to the command of the pain, which tells me to protect my leg by keeping it still. Note the different goals of the two motivational states—the latter motivates me to protect a part of my body, whereas the former motivates actions that will have an effect on one of my sensations.

Some secondary motivation occurs in virtue of our beliefs about our pains. Because I have a toothache, I might come to believe that I have a cavity. I also believe that if I have a cavity, I should go to the dentist. My pain thus motivates me to go to the dentist. Here, what motivates is *the fact that I am*

1 I owe the distinction and the terminology to Price (2000), although note that I'm carving it differently than he does. Note also that I'm making the distinction in a different way than I did in Klein (2007). There, I confused primary motivational force and felt unpleasantness; the latter I now include in secondary motivation.

in pain along with my appreciation of what that entails. That motivation is clearly contingent: my toothache would not motivate me in this way if I didn't have any beliefs about cavities. Conversely, I need not actually *be* in pain to be motivated in this way. My toothache yesterday is just as good a reason to go to the dentist today, even if it has subsided. The same sort of motivation can occur even when it's not *my* pain that's at issue: the fact that my child has a toothache should motivate me, by roughly the same reasoning, to take him or her to the dentist.

Similarly, pains might cause emotions, which are themselves motivating states. Pain might be an occasion for anxiety, for fear, for vigilance, or for anger. These emotions are, again, often highly contingent on my other mental states. The same twinge in my arm that I previously ignored might now be an occasion for fear after a diagnosis of heart disease.

Not all pains provoke these secondary emotions. Further, these are the same sort of emotional reactions that can occur in response to other cues and so are obviously distinct from the pain. Many forms of secondary motivation are thus obviously extrinsic to pain and only contingently connected to it.

There is a more complicated case, however. One important kind of secondary motivation arises in virtue of the way pains *hurt* or cause us to *suffer* or are *painful* or *feel bad*. I'll use these terms to pick out the same phenomenal character felt by many pains, a character that is disagreeable and dislikable. When a pain hurts, we gain a motivation to end that very pain. That motivation might outstrip, in an important sense, the primary motivation of a particular pain. Because my pain hurts, I go to a doctor, take medicine, and whinge. That's more than my pain tells me to do.

That a pain hurts is often the most salient fact about it. The fact that pains make us suffer plays a large part in explaining why we avoid pains. I might avoid going to the dentist because I know he has a free hand with the drill, and that will hurt. Again, what motivates me here is that I expect to be in pain, and that pain is bad and something I want to avoid. When these negative emotions are present, I suspect that they are primarily reactions to the *painfulness* of pain rather than to pain itself.

These secondary reactions to pain are important and often more pressing than the primary motivational force of pain. To call a bit of motivation secondary, then, is not to denigrate it: a lot of what pain motivates, it motivates in virtue of suffering.

I claim, however, that suffering is not a feature of pain: it is a response *to* pain. This means that suffering is only contingently connected to pain, and hence that pains only contingently hurt and feel bad. That is a controversial thesis. You will probably want some convincing. This chapter will argue for the distinction between pains, on the one hand, and hurt/suffering/painfulness, on the other. Distinguishing suffering and pain makes imperativism more plausible. Imperativism, recall, was motivated by a specific view about pain's biological role and its similarity to other states. Those other states are obviously only contingently connected to suffering—one can be so hungry that it hurts, but no one thinks that hunger feels bad intrinsically. I claim that suffering and pain, while more *frequently* connected, are similarly only contingently related.

Having made the distinction, the imperativist faces two different and distinct explanatory tasks. Primary motivational force is key for the imperativist. The whole point of pains is to motivate. So the imperativist must say something about how and why imperative sensations motivate those who have them.

Secondary motivation, in contrast, requires a different explanatory story. Because secondary motivation is an extrinsic and contingent fact about pains, the imperativist does not strictly speaking owe a story about what suffering is. At most, I owe a story about *why* we suffer—that is, why we have thus-and-such reaction to pains and not some other. (Compare: it is an extrinsic fact about crocodiles that they frighten me. Someone giving an account of what a crocodile is might, ideally, say something that explains why they are so frightening. That doesn't require giving a story about what *fright* is, however—they might point out that crocodiles are big and strong and dangerous and full of teeth, and that most such things are frightening, so it's really no surprise, you know?)

The same is true, note, of all secondary reactions. Pain sometimes makes us anxious. As an imperativist, I think there's a straightforward story to be told about why pains make us anxious. Roughly, pains limit motion, and things that limit motion tend to produce anxiety because they increase the chances that bad things might happen to us. This story doesn't require explaining what anxiety *is*: it merely makes a case that pains are among the anxiety-producing things. That's the same thing that the imperativist needs to do with suffering.

I'll return to this question again in chapter 14, where I will take up the second part of this explanatory task. That is, having established that suffering is a contingent feature of pain, I'll explain why pains hurt. Nowhere will I fully explain what suffering is, but I will gesture at some plausible stories.

For this is ultimately a book setting out a theory of *pain*, not about things only contingently connected to pain. Suffering is of interest only because imperativism can say something novel about why pains cause us to suffer. The sense in which pains have intrinsic motivational force is due solely to their imperative nature. It is that primary sense that I will be most keen to explain and deploy in explanations.

4.2 Pain and Suffering

Most pains feel bad. They feel bad in a particular way: they *hurt*. They are *painful*. They cause you to *suffer*. I'll take these three terms to pick out a quality shared by most pains. Sensations that hurt have a distinctive phenomenology.[2] Suffering is a more specific way in which something can feel bad or unpleasant. The sight of leeches attached to your leg is disgusting and so deeply unpleasant, but it does not hurt: it feels bad in some other way.

The phenomenology of suffering is strongly motivating. Because pains hurt, we generally want to avoid them and get rid of them when we feel them. Insofar as pain is a bad thing (and unnecessary pain an uncontroversially bad one), it is because it causes suffering. Because hurt is an extremely salient feature of most pains, many have assumed that it must be a constitutive, intrinsic, or necessary property of pain.[3]

An imperativist should be wary of such claims. Pure imperativism must reject them outright. Most pains do feel bad, to be sure, but the primary motivational force of pain—and the one that is wholly constitutive of pain on a pure imperativist account—is simply a command from the body to protect the affected body part. Commands from the body do not intrinsically, essentially, or necessarily feel bad. I have argued that pain is a homeostatic sensation and so comparable to hunger, thirst, and cold. But the mild forms of these sensations do not feel bad. Mild hunger is a command to eat and motivates me as such. But that doesn't feel bad or good: it's just a sensation, with no particular

2 The term "hurt" is not ideal. It is ambiguous between a phenomenologically rich mental state and the physical cause of that state ("I hurt my foot by stepping on a nail"). I will use it in the former sense. It is also ambiguous between the phenomenologically rich affective reaction we have to states like pain and the disposition that pain has to cause such reactions. I'll again use "hurt" in the former sense and talk about *why* something hurts when talking about the latter. "Painfulness" suffers from a similar ambiguity as "hurt," however. More importantly, "painfulness" has been co-opted for a different purpose by Nikola Grahek. Grahek (2007) argues for a composite theory; he claims that pains have a non-motivational sensory aspect, which he calls "pain," and an affective aspect, which he calls "painfulness." I deny the composite view of pain. Imperativism says that pains consist of a single, non-motivational sensory state constituted by an imperative; insofar as pains hurt, that is an extrinsic, contingent fact about them.

3 For a nice review, focusing on the purported intrinsic badness of pain, see Swenson (2006).

further valence. So the pure imperativist should also avoid saying that pains feel bad intrinsically, necessarily, or essentially.

As ever, I intend to make a virtue out of apparent vice. For *nobody* should say that pains must feel bad. Doing so, I'll argue, relies on a confusion between pain and a property shared by pains and many other sensations. To that end, I'll give several arguments for distinguishing between pain and suffering.

4.3 Arguments for the Distinction

4.3.1 The Argument from Dissociation

The simplest argument for distinguishing between pain and hurt is this: they come apart. Some pains don't hurt, and many things that hurt aren't pains. The two states thus dissociate: it is possible to have each without the other. So we have good reason to distinguish the two phenomenal qualities.

Begin with the question of whether pains always hurt. Many pains clearly do, and when we think about pains, we understandably recall those that feel bad. Nevertheless, I think there are pains that don't hurt, that we don't have any obvious negative reaction to, and that don't feel bad.

The point is often made by reference to morphine pain, pain asymbolia, or similar pathological cases (see e.g., Hall, 1989). These present interpretive difficulties, so I'll discuss them further in chapters 11 and 12. The point does not rest on such *recherché* cases, however. Mild pains provide simple, if less striking, examples. As R.M. Hare (1964) puts it:

> There are, in fact, small degrees of pain which are by no means disliked by everybody. Most people could draw the point of a needle rather gently across their skin (as in acupuncture) and say truthfully that they could distinctly feel pain, but that they did not dislike it. Some might say that they would rather be without it than with it; but that would apply to a great many sensations about which no philosopher, to my knowledge, takes the line that some do with pain. Most people would rather be without a feeling of giddiness (though children often induce it in themselves out of interest); but nobody says that no sense can be given to the sentence "I feel giddy, but do not dislike it." (97)

I think Hare is right. Mild pains in uncomplicated circumstances don't really feel bad. Here's another, non–self-inflicted example. The pains that precede unproblematic postural adjustment are recognizably pains. In ordinary cases, however, they don't really hurt. They motivate you to change your position. That motivation proceeds without a hitch, and so the pains are hardly worth noticing. Of course, these pains might *become* painful—transpacific flights are an exercise in suffering from postural pains. But these needn't hurt, and normally they don't.

Sustained introspection throughout the day reveals that many pains—while still motivating—aren't unpleasant. (I urge this sort of introspection as an exercise on the reader.) At the least, it becomes hard to tell whether, for example, a brief and easily eliminated postural cramp really hurts. But if hurt were an essential, intrinsic part of pain, then it would be odd that we could be in doubt about whether a particular pain hurts.

In the other direction, it's clear that pains are not the only mental states that hurt. Many sensations can hurt—they can be *painful*—when intense. Pressure, stretching, itch, and thermal sensations (considered as both tactile sensations and the distinct sensations associated with core temperature) can all hurt. Other interoceptive bodily sensations can be painful too: you can be painfully hungry or painfully tired, for example. Dyspnea is a particularly painful sensation if it persists. Even ordinary representational sensations can become painful: a concert can be painfully loud, the sun painfully bright. Describing a familiar sensation, Erasmus Darwin (1918) describes, "The disagreeable sensation called the tooth-edge is originally excited by the painful jarring of the teeth in biting the edge of the glass, or porcelain cup," and notes that this painful sensation is subsequently "excitable not only by a repetition of the sound, that was then produced, but by imagination alone" (14). The variety of physical sensations that can be painful seems almost unlimited. As W. Noordenbos noted, "pain may arise from virtually *any* type of stimulus or may be the result of afferent patterns which may travel via *any* available pathway" (quoted in Bakan, 1968, 63).

Hurt is not limited to physical sensations. Many emotions hurt. Grief and heartbreak are notoriously painful. In his discussion of anger in the *Rhetoric*, Aristotle defines anger as "an impulse, accompanied by pain, to conspicuous revenge for a conspicuous slight" (*Rhetoric* 1378a31). "Pain" is a translation of the Greek *lupē*. As Cooper (1996) notes, *lupē* is applied to both ordinary bodily pains and painful emotional states (in both Aristotle and in non-philosophical Greek); it has a special connection to the anguish of grief in poetic contexts (245). In summary, not all pains hurt, and not all things that hurt are pains. They come apart even in everyday cases. That's a good reason to distinguish between pain and suffering.

Note, as a final consideration, that there is an exactly parallel story about the *pleasant*. Many different things are pleasant. Conan found it pleasant to crush his enemies, to see them driven before him, and to hear the lamentations of their women. That pleasure qualifies three different states—two distinct sensory modalities, and one more intellectual appreciation of the world. Of course, what is pleasant depends partly on context and what you value. For us non-barbarians, different things are pleasant: a hot bath, a kiss, a beautiful painting, or hard intellectual work. In the case of pleasure, the distinction is clearer because we don't really have a word, as we do with pain, to mark a pure bodily sensation that is prototypically pleasant. Pleasures are too diverse and idiosyncratic for any particular one to get the honorific (although the slang expression "orgasmic" is applied rather widely to very pleasurable sensations, and that might be a rough parallel).

4.3.2 The Argument from Independent Variation

I find the argument from dissociation compelling, but there are places you might balk. Appeal to individual experience always carries a risk, particularly when it's the appeal to the *absence* of a particular phenomenology. Why not think, for example, that very mild pains hurt but simply don't hurt that *much*? If the suffering associated with mild pains is present but minor, then perhaps pains always come with suffering after all.[4]

This objection, however, already contains within it the core of a further argument in my favor. First, remember that the question at hand is whether pain and suffering should be distinguished as distinct phenomena. It is *not* the question of whether pain can occur without associated suffering—I think it can, and that this provides the clearest reason for distinguishing pain and suffering, but one could maintain the distinction without believing that they actually come apart. Second, note that the objection relies on the possibility that *suffering* comes in degrees as well—it can be so mild as to be unnoticeable. Third, if pain and suffering are *not* distinguished, then pain and suffering should *covary* in their intensity.[5]

Yet pain and suffering are able to vary independently in their intensity. On the one hand, mild pains can cause intense suffering—as when, for example,

4 Thanks to an anonymous reviewer for pressing this objection.

5 I say "covary in" rather than "be identical with" for the sake of charity to my interlocutor. For if the relationship between the two intensities was identity, then one would need a story about why one could notice mild pain without mild suffering. Covariance is a looser relationship and so allows attention and other cognitive states to modulate suffering *judgments*. Any modulating factor, though, should probably be limited to a relatively simple linear transformation from one intensity domain to the other—more complex relationships are hard to square with identity.

they are not under your control, they forebode serious injury, or they have persisted for weeks. On the other hand, repeated pains can cause one to suffer less and less. A new exercise routine or getting blood drawn might become less and less unpleasant over time. If you must have blood drawn every day, you'll find it less and less unpleasant. It's not so much that the pain gets less intense, however: there's still the sting of the needle, and that doesn't change much at all over time. Rather, as you become familiar with the same pain, it *hurts* less. You don't mind it so much. It is pain, but it doesn't feel as bad as it did when you started—perhaps because it's familiar, perhaps because your anxiety has diminished, perhaps because you've learned that nothing bad will come of the pain. When that happens, you discover that the pain isn't all that bad. Finally—and details will be examined in chapter 12—some painkillers such as Vicodin appear to work primarily by reducing suffering rather than the intensity of pain.

Of course, most intense pains are intensely unpleasant and mild pains less so. That's a fact that any theory about the relationship between pain and suffering should account for. But the fact that pain intensity and intensity of suffering can co-vary suggests that they are two distinct phenomena. Further, I think we can do a bit better than introspection in this case. In a well-known study, Rainville et al. (1999) found that "pain intensity" and "pain unpleasantness" ratings could be independently modulated by hypnotic suggestion. These findings are often interpreted within a framework in which both dimensions are constitutive of pain. It seems to me, however, that they are equally good evidence for the view I propose—that is, the view on which pain intensity alone is part of pain, while unpleasantness is a secondary characteristic.

4.3.3 The Argument from Differing Domains

To deny the distinction between pain and hurt is to claim that all things that hurt really deserve the name "pain." Heartbreak is, properly speaking, a pain of exactly the same class as a stubbed toe. As an example of such a view, one might argue that "pain" is just a property that qualifies any strong negative sensation (Hare attributes such a view to Kurt Baier; see 1964, 29ff). "Pain" and "hurt" would then be rough synonyms, picking out the same class of sensations. In ordinary speech, we often don't distinguish the two words, which provides some evidence for this identification.

I think such an identification is implausible for a number of reasons. For starters, note that there are two classes of pains on such an account. Some pains qualify another, distinct sensation—a tactile sensation, if nothing else.[6] But not all pains seem to have a kernel of some other sensation, not even a tactile core. Sometimes my ankle merely hurts, with no other accompanying sensation. So some pains qualify sensations whereas others don't. That is at least odd: despite feeling the same, sometimes pain ends up as a property that qualifies another sensation, and sometimes it's just freestanding. Far more natural, I submit, is simply to distinguish two states: pain, which is a freestanding sensation in its own right, and hurt, which always qualifies some other mental state.

Further, there seem to be two different motivational states associated with painful pains, each with a distinct end. On the one hand, bodily pains motivate actions designed to protect bodily integrity: the pain of my sprained ankle motivates me to avoid walking on it, the pain of my burning hand to pull back from the flame, and so on. On the other hand, pains—and all painful sensations—give us motivation to *remove the sensation*. But that second motivation needn't coincide with the first, body-oriented motivation. Indeed, it is often orthogonal to it. Because the pain in my ankle hurts, I have a motivation to take powerful narcotics. Those would also remove the pain, however, which would make it more likely that I would walk on my ankle and so more likely that I will injure it further.

Indeed, the two come apart in a familiar setting. Suppose I bark my shin. A natural response is to rub it, which will reduce the pain I feel by lateral inhibition from nearby wide dynamic range neurons (Melzack and Wall, 1996). Rubbing does nothing at all to help my shin and has little to do with protecting it. Instead, it is a trick we all learn to get rid of certain kinds of pain. That such rubbing works is contingent on our peripheral physiology and has nothing at all to do with pain's basic biological function.

The point is not just to emphasize the oddity of a single state motivating two actions with opposing ends (although that *would* be odd and demand a story). The point is rather that the two kinds of motivation have wholly different targets: one motivates actions with respect to the *body*, whereas the other motivates action with respect to a *mental state*.

We find a similar structure when we examine other painful states. There, the opposing nature is more obvious. Heartbreak motivates various actions: to find another love, say, or to try to win her back. Because heartbreak is a painful feeling, however, one can also be motivated to eliminate the feeling of

6 I assume that one cannot merely conjoin pain and some sensation ϕ to become painfully ϕ. If I'm hungry and you punch me in the stomach, I don't thereby become painfully hungry.

heartbreak itself: to drink to forget, to lose oneself in meaningless sex, or to join the Foreign Legion. The latter actions might eliminate the feeling of heartbreak without achieving the ends that heartbreak actually motivates (indeed, they probably work against it). Here the distinction is obvious: heartbreak motivates actions that have goals in the world, whereas the painfulness of heartbreak motivates actions with goals directed toward one's own sensation. The same is true of pain and suffering.

4.3.4 The Argument from Common Phenomenology

A final argument for distinguishing pain and hurt is that if we don't, we miss out on the distinctive phenomenology of pain. For the things that *hurt* are extremely heterogenous.[7] It is hard to see what they might have in common except that they hurt.

For example, Korsgaard (1996)—who has many insightful things to say about painfulness, hurt, and feeling bad—claims:

> If the painfulness of pain rested in the character of the sensations rather than in our tendency to revolt against them, our belief that physical pain has something in common with grief, rage and disappointment would be inexplicable. For that matter, what physical pains have in common with each other would be inexplicable, for the sensations are of many different kinds. What do nausea, migraine, menstrual cramps, pinpricks, and pinches have in common, that makes us call them all pains? (148)

The first sentence is completely reasonable, and the second baffling. The class of "physical pains," as Korsgaard calls them, obviously have something in common. Presumably that common feel is precisely why everyone can readily distinguish them as pains. As Hare (1964) puts it,

> It seems, rather, that there is a phenomenologically distinct sensation or group of sensations which we have when we are in pain, and that there could be (whether there actually is or not) a word for this group of sensations which did not imply dislike. I say "group of sensations" because burning pains, stinging pains, stabbing pains, aches, *etc.*, are distinguishable from one another, although they clearly fall into a group which is bound together by more than the fact that they are all disliked. (94)

Of course, it's hard to *say* what that common thing is other than that each is a pain. That's true of every sensation, however; nobody objects to the idea that there are blue sensations because it's hard to put the differences in shades of

7 A similar problem exists for pleasure; see Smuts (2011) for a nice review of the objection from heterogeneity.

blue into words. The distinction is readily made in ordinary speech when peo-ple distinguish "physical pain" from other sorts of painful sensations. Further, as I emphasized in chapter 3, there are obvious biological advantages to having a sensation that gets you to protect your body in response to physical threat.

In summary, separating pain and hurt takes an apparently heterogenous phenomenological set and distinguishes two phenomenologically homogenous partitions. There's pain, which is a distinctive sensation with a distinctive motivational role, and there's hurt, which qualifies an extraordinarily heterogenous set of mental states but promotes the same activity (something like dislike and the motivation to eliminate) toward each of the token mental states that it qualifies.

4.4 Recursion, Not Regress

One potential worry about the distinction between pain and hurt must be addressed. I claim that pains hurt, and that is the way in which they feel bad. But surely hurting is a way of feeling bad as well? That is, it doesn't just feel bad to be in pain but also to hurt. But if it hurts to hurt, then surely that also hurts, and a regress looms. A papercut hurts, but it doesn't bring an infinite tower of hurt in its wake.

This objection misunderstands the relationship between hurt and pain. Insofar as pains hurt, they feel bad. They feel bad for various reasons. Some of those I'll cash out in chapter 14. But one reason why they feel bad is just that they hurt. It is an objectionable fact about most pains that they hurt, and that is in turn one of the reasons why they hurt. As Korsgaard (1996) puts it "An animal which is in pain is objecting to its condition. But it also objects to being in a condition to which it objects" (154).[8]

This implies neither regress nor logical circle, however. Rather, it involves what Korsgaard calls a "recursive structure" that makes use of unproblematic self-reference to make a property of a whole depend constitutively on the instantiation of that property.

Compare: I go to a dull movie. I dislike it. One of the things I dislike about the whole experience is that I wasted my time on a dull movie when I could have gone to the beach instead. So one of the reasons why I dislike the

8 Note that in this passage and surrounds, Korsgaard uses "pain" to refer to what I'm calling "hurt."

movie involves the fact that I disliked the movie (for it is only if I disliked the movie that I wasted my time). But there is no logical circle here, and hence no philosophically troubling regress.

What might occasionally happen is a particularly nasty *practical* regress: I might resent the movie for having wasted my time, including the time that I'm wasting thinking about resenting the movie. That can quickly spiral into a black mood. Pains can do that too: I can begin to worry about my pain and then worry about worrying about pain, and that can spiral quickly into an unpleasant state, far worse than the original pain.

I think this might explain some aspects of so-called *pain catastrophizing*. The pain catastrophizing scale (Sullivan et al., 1995) contains many descriptors gauging the subject's negative reactions to the negative aspects of their pain. Catastrophizing is linked to increased pain and worse outcomes (Sullivan et al., 2001), suggesting that it is possible to fall into a circle like this that enhances the negative aspects of pain. But because catastrophizing varies among people and is generally a contingent matter, it also suggests that any potential circle is a practical one not a logical one.

4.5 Conclusion

This book is about pain, not suffering. I have no theory of suffering, no story about what sort of state is present when pains feel bad. I've already claimed that suffering is a contingent feature of pain, and that pains which don't cause suffering, while comparatively rare, are perfectly intelligible. So suffering isn't constitutive of pain, nor is it a necessary feature of pains. A philosophical theory about what makes pains *pains* needn't include a theory about suffering.

Of course, it is an important fact about pains that they *do* cause suffering, and paradigmatically so. Part of the reason why "pain" is ambiguous is just that pains are the most common sensations that hurt; "pain" thus becomes a synecdoche for the class of sensations that cause suffering. A good theory of pain should say something about *why* pains typically hurt. I return to that project in chapter 14, along with the more complicated question of whether hurt is a reaction to intrinsic properties of pain or whether it has some more flexible relationship. Before we get there, imperativism will need further fleshing out and defense.

5 The General Content of Pains

5.1 The Content of Pain

I claimed in chapter 3 that pains were protection imperatives. It is time to expand and defend that claim.

At heart, a protection imperative is a command to protect part of your body. It is satisfied just in case you actually protect the part of your body. The content of pains, at a first pass, is thus a command to protect a particular body part in a particular way with some level of urgency. The content of each pain can be expressed as an instance of the following schema:

PS: Keep *B* from *E* (with priority *P*)!

B stands for a particular body part, *E* a nominalized passive gerund phrase, and *P* a ranking function. I will discuss each of these in detail in chapters 7 and 8. I claim that each pain has a content that is an instance of schema **PS**. (Sometimes I'll speak simply of commands to protect, but these should always be understood as implicit commands to protect a certain part, in a certain way, with a certain urgency.) So, for example, the pain of a broken ankle is the command, "Keep your ankle from bearing weight (with urgency so-and-so)!" The pain of a burn on your arm is, "Keep your arm from being touched!" The pain of a torn biceps muscle is 'Keep your biceps from contracting!' And so on.

Each of these commands promotes certain kinds of action that, under normal circumstances, fixes the problem eventually. Crucially, **PS** is meant to be a schema for *content*-bearing states such as pain. Mere reflexes might also motivate you to protect a body part in a way that pains would—for example, by withdrawing from a flame. I take it that reflexes are not mediated by sensations, however, while pains are interesting because of their role in promoting behavioral homeostasis through a particular sort of content-bearing homeostatic sensation. Schema **PS** is meant to give an account of that content.

I begin by discussing imperative content generally. The details are important because they explicate imperativism and can be deployed to solve various puzzles. Then I will talk a bit more about the advantages of protection imperatives as against other imperativist proposals for the basic content of pain.

5.2 Imperative Content

5.2.1 Some Basic Ideas

Ordinary indicatives express a proposition and are meant to correspond to the way the world is. Imperatives are different. One does not utter an imperative to express something true but to bring about a state of affairs. Thus, imperatives express *commands* rather than propositions. In virtue of these different contents, imperatives and indicatives differ in their direction of fit (Searle, 1985).

Indicatives are truth-apt and are made true or false by the way the world is. Imperatives, by contrast, are *satisfied* if the world is made to be a certain way and *unsatisfied* if not. An imperative is satisfied just in case the addressee brings about the action enjoined by the command that the imperative expresses. Satisfaction is plausibly extensional: the command, "Bring me a soda!" is satisfied just in case you bring me a soda and unsatisfied if you don't.

Imperatives may also be *legitimate* or *illegitimate*. You might reasonably protest that I don't have the authority to order you around. The legitimacy of orders is determined by an explicit set of rules in some contexts; in others, we rely on looser guidelines. While the legitimacy of a command may depend on its content, legitimacy is orthogonal to satisfaction. Illegitimate commands might still be satisfied, and legitimate ones go unsatisfied.

Facts about legitimacy do not seem to be part of the content of imperatives. For one, we might wonder whether a given command is legitimate ("Can he really order me to change my seat?") while still understanding the content of the imperative issued. For another, the same imperative can vary in legitimacy depending on who issues it. The command expressed by, "Show your ticket!" is legitimate if the usher says it and illegitimate if I say it, yet we have expressed the same command. Finally, we are sometimes forced to make the source of legitimacy explicit when we issue commands ("I'm your father, and I say don't see that boy again!"). This would be redundant if facts about legitimacy were part of imperative content. It is not redundant, however, because an imperative inherits its legitimacy from extrinsic facts about the relationship between the issuer and the addressee. Those facts might need emphasizing because they are not part of the content.

This suggests that the content of imperatives can be more or less identified with their satisfaction conditions alone (although this simple picture will need [illegible]). Here, I am guided by Hamblin's (1987) semantics for imperatives, especially his *addressee-action-reduction* theory of imperative content.

The basic idea is simple. Imperative satisfaction is an extensional notion. The imperatives:

> Bring me a soda!
> Find the bottle opener within the next five minutes!
> Don't let the fridge go empty!

are satisfied just in case I, respectively, bring you a soda, find the bottle opener, and keep the fridge from going empty. At a first pass, let us identify the content of an imperative i with its satisfaction conditions, modeled as the set of possible worlds W_i that share the features commanded.[1] The content of the first imperative is just the set of possible worlds in which you bring me a soda and similarly for the other two.

5.2.2 Further Details

This first pass will need refinement. For starters, imperatives really enjoin *actions* not states. This is explicit in the first imperative and implicit in the second and third. I don't really satisfy the second imperative if I just stumble over the bottle opener: the command tells me to *look* for the opener, not merely to end up finding it. Even apparently "stative" imperatives like the third really enjoin the addressee to take or avoid actions that would *bring about that state* (Hamblin, 1987).

Similarly, imperatives not only enjoin me to take (or avoid) certain actions but also to take any reasonable steps necessary to perform (or avoid performing) the explicitly enjoined action. If I fail to get you a soda, then I don't excuse myself by noting that I can't reach one from my chair. Imperatives demand that I undertake a series of steps that will lead to the specified outcome. Most imperatives are open ended—they don't demand that I take any *particular* course as long as the course terminates in the actions explicitly enjoined. So let W_i be restricted to the set of worlds in which I undertake a

1 A standard caveat about model-building. I am not suggesting that your head somehow *literally* contains sets of possible worlds. This is intended to be an abstract, partially idealized model of the content of imperatives. It is justified to the extent that it allows us to capture philosophically interesting features of imperatives. There will be a further story to be told about how this content is embodied in actual perceivers, how the content so embodied is causally effective, and so on. I leave that to philosophers who specialize in the embodiment of content. See especially Martínez (2010) for an account tailored to imperativism.

series of actions (“strategies” in Hamblin’s terminology) that are likely to lead me to the enjoined action.

Note that considerations of strategies (rather than merely endpoints) also help us make sense of the important difference between *trying* to satisfy an imperative and failing versus merely failing to do anything, although both worlds might not be in W_i. This will end up an important consideration. For I will argue later that imperatives can be unsatisfiable while still motivating. A natural way to cash this out is in terms of strategies—strategies that make an imperative more likely to be satisfied or would satisfy the imperative in closely related worlds.

Further refinements to this story are possible. Some of these are general and designed to capture features of all ordinary language imperatives. One might, as per Hamblin (1987), include considerations of time and agent in the imperative and restrict W_i worlds to ones where the satisfaction of the imperative is after the time of utterance. Vranas (2008) argues that we need both satisfaction conditions and violation conditions, in part, to account for imperatives that are neither satisfied nor violated. Many of the complications arise when one tries to account for imperative conditionals and imperative inference more generally (an issue that goes back at least to Ross, 1944; see Parsons, 2012b; Vranas, 2010, for recent review and arguments).

5.2.3 The Content of Protection Imperatives

Other refinements are necessary to deal with the restricted class of imperatives that will be used to analyze pain. I have argued that pains are commands to protect a certain body part. Protection imperatives have several important features.

One can protect a body part in two senses. On the one hand, I protect my ankle by not walking on it—that is, by avoiding certain kinds of positive action myself. On the other hand, I protect my ankle by keeping things from coming into contact with it. That might involve avoiding kickball games, fending off children and dogs, and so on.

Call these two senses the *active* versus *passive* ways of protecting something. Active and passive protection require commands with slightly different satisfaction conditions: active protection commands are satisfied when you do something, whereas passive ones are satisfied if you *refrain* from doing something.

Most pains command a mix of active and passive protection. Depending on the situation, one or the other might dominate. A sprained ankle generally calls for active protection: the most important thing is to avoid walking and other weight-bearing activities. A burn on my arm, by contrast, mostly calls for passive protection: I need to keep things from hitting the burned patch so it can heal. I'll mostly collapse the distinction between active and passive protection in what follows, although I will have a bit more to say about it in chapter 7.

A further cut is necessary. Some imperatives are discharged as soon as they are satisfied. If I command you to bring me a soda, then you're done once you bring it. *Standing* imperatives, by contrast, remain in force even as one complies with them. If I tell you to keep an eye out for wolves, then I do not want you to look for a bit and then pop off for a drink: I want you to look now and keep looking. Protection imperatives should be understood as standing imperatives. The command to protect your ankle remains in force even as you comply with it. That's important because pains need to limit action not only now but also in the future. I can't rely on my ankle and so shouldn't plan to use it in the near future.

Some standing imperatives are implicitly unlimited in duration ("Do a good deed daily!"). More common, however, are standing imperatives of limited duration—they have either a definite endpoint ("Look for wolves for the next five hours!") or an indefinite conditional one ("Look for wolves until I say otherwise!").

I suggest that pains are indefinite conditional standing imperatives. The pain in my ankle orders me to protect my ankle until the pain ceases. It is an imperative that remains in force as long as it continues to be issued and until it stops. In that sense, pain is again like a fire alarm. The fire alarm gets you to evacuate, and stay evacuated, for as long as it rings. So the condition on the standing imperative is simply the continuation of pain. Of course, some pains come and go quickly. Brief pains may be treated simply as standing imperatives that are retracted shortly after they are issued. (Entirely possible in everyday speech: "Watch out for wolves! Wait, wolves—run!")

The role of pain in limiting future actions is often overlooked. Yet it is a crucial part of pain's overall function. If I am injured in a way that proscribes against certain kinds of actions, then it would be bad if I relied on those actions as I planned my day. It's no use trying to protect me if I inevitably get in situations where I'll have to walk anyway. Rather than acute pain reminding me that my ankle is broken every time I try to walk, I have a standing imperative against walking that increases in intensity if I neglect it. That keeps me from walking now and from knowingly getting into situations where I would have to walk.

5.3 Imperatives and Ranking Worlds

A final consideration will be relevant when it comes to talking about the intensity of pain.[2] I suggested that the content of imperatives could be identified with a set of *satisfaction worlds* W_i. This still leaves something out. Two commands can be satisfied in the same worlds and yet differ in subtly different ways.

Suppose I give you the command:

[O]: Raise money for Oxfam!

[O] admits of two readings: one on which you are commanded to raise *any* amount of money, another on which the more money you raise the better. These are two different commands. Both imperatives would be satisfied in the same worlds: namely, ones where I raise any amount of money for Oxfam. Yet they are clearly different commands. This is because the first is indifferent among the W_i worlds, whereas the second *ranks* W_i worlds: the ones where I raise more money are more preferable. What distinguishes these two readings must be some difference in how they rank worlds.

So rankings of different outcomes must be, at least sometimes, part of the content of imperatives. Indeed, even ordinary imperatives can be treated as a special case of these more complex ones. The default ranking, as it were, is one where each satisfaction world is ranked equally high and strictly better than any non-satisfaction world.

I propose, then, to identify the content of an imperative with the pair $\langle W_i, \gtrsim \rangle$. W_i is the set of satisfaction worlds, whereas $\gtrsim$ is a ranking function defined over both satisfaction and non-satisfaction worlds. The function itself ranks pairs of worlds and introduces a partial ordering over the set of all worlds (i.e., over both satisfaction worlds and non-satisfaction worlds). For any two possible worlds w_i and w_j, $w_i \gtrsim w_j$ means that w_i is ranked at least as high as w_j. Two worlds are equally ranked if they are ranked at least as high as each other, and a world w_i is ranked strictly better than w_j just in case w_i is ranked at least as high as w_j and they are not equally ranked.

2 The material in this section derives from joint work with Manolo Martínez.

For the simplest sorts of imperatives, each satisfaction world is ranked strictly better than each non-satisfaction world, each pair of satisfaction worlds are equally ranked, and each pair of non-satisfaction worlds is equally ranked. That is, $\succsim$ partitions the entire space of worlds W into exactly two equivalence classes: the satisfaction worlds W_i and the rest, with worlds in the former preferable to the latter. Complex imperatives use $\succsim$ to partition W into more fine-grained equivalence classes.

This is the case with **[O]**. On the first reading of **[O]**, $\succsim$ simply defines a partition with two equivalence classes: one that includes all and only those worlds in which the addressee raises any amount of money for Oxfam, a second that includes every world where the addressee raises no money at all, and ranked such that any world in the former equivalence class is strictly preferable to any world in the latter. On the second reading of **[O]**, $\succsim$ will provide a more complex ranking—the satisfaction worlds will be partitioned into smaller equivalence classes, each composed by all and only those worlds in which the addressee raises a certain particular amount of money, and ranked such that worlds with more money are preferable to those with less. The distinction between the two meanings of **[O]** can thus be grounded in the differences in the ranking function of the two readings not in the satisfaction conditions. That again suggests that $\succsim$ is part of the content of imperatives and necessary to make certain distinctions among them.

Because $\succsim$ is defined over non-satisfaction worlds, imperatives can also rank worlds in which they *fail* to be satisfied. The sergeant orders the private to peel a truckload of potatoes, knowing that success is unlikely. But implicit in his order is a ranking of worlds in which the private fails, something like, "If you can't peel all of the potatoes, then you'd better peel as many as you possibly can!" Hamblin (1987) notes that these secondary commands are sometimes explicit, as in, "Come to the meeting; if you don't, at least send an apology" (84).[3]

Finally, some complex commands might actually rank non-satisfaction worlds as preferable to some satisfaction worlds. Suppose my wife tells me to kill the cockroach in the bathroom. I pour some gasoline, toss in a match, and thereby satisfy her command. As we stand outside the charred ruins, she might reasonably protest that I've misunderstood her command: she did not

3 To be clear, however, Hamblin seems inclined to treat this as a case of complex conditional content, and his account has no mechanism for ranking satisfaction worlds. I think the use of a ranking function is a more general strategy than appeal to conditional imperatives.

intend me to kill the cockroach *at all costs*, but in some manner that kept everything else roughly as it was. One way to handle these complex contents that preserves the surface grammar of the command is to appeal to the ranking function: $\gtrsim$ ranks some satisfaction worlds (the ones where I did what I did) *lower* than some non-satisfaction worlds (the ones where the cockroach lives), although of course some satisfaction worlds (ones where I use a shoe) are ranked higher still.

This is not the only way to treat complex commands, of course: one might treat the implied satisfaction conditions as indefinitely complex ("Kill the cockroach *and* don't burn down the house *and* don't poison us all *and...*"), and in many cases, that complexity can be represented by something analogous to a *ceteris paribus* clause. But given that we need something like a ranking function anyway, we can use it to preserve a simple relationship between the spoken imperative and the satisfaction conditions and account for the complexity using that ranking.

The relationship between rankings and imperative intensity will require a chapter of its own to untangle; I will handle that in chapter 8. For now, I will treat imperative contents as an ordered pair $\langle W_i, \gtrsim \rangle$. When the ranking isn't relevant, I will sometimes simplify by talking just about the satisfaction conditions of the imperative. Even with those simple imperatives, a ranking function should always be assumed to be present (and typically one that ranks W_i worlds above all others).

5.4 Protection Imperatives

I claimed that pains are best understood as protection imperatives. Other imperativist accounts exist. Each of them is similar to the above account in broad outline. Each gives a slightly different story about the positive imperative content of pain. These alternatives coincide with protection imperatives in ordinary cases. The choice among the various imperative accounts is not obvious. So I will defend and elaborate my account by arguing for protection imperatives rather than some other plausible alternatives.

5.4.1 Stop Imperatives

First, pains might simply say "Stop!" That is, pains might be imperatives against taking some action that you're actually taking. So, for example, Richard Hall (2008) writes that:

> ... with some pains there is also an imperative component. Roughly: "Stop! Stop doing what you're doing with this bodily part!" The sharp pain in your lower back when you're lifting something heavy tells you to stop doing that. The pain in your hand when you grasp something hot or sharp tells you to stop doing that and withdraw. (534)

Stop imperatives fit well with acute pains. Because many ways of protecting your body involve ceasing to do what you're doing, stop imperatives will often end up extensionally equivalent to protection imperatives. Hall believes that pains are composites of both imperative and representational content. The suggestion is available to both pure and hybrid accounts, however.

That said, I think pains can't be stop imperatives. Stop imperatives don't fit well with pains of recovery. As I lay on the couch with my sprained ankle, there's nothing I can stop doing. Nevertheless, my ankle hurts. Why? Because I need to avoid *planning* to do anything with my ankle, and because I should be vigilant for things that might come along and hurt my ankle more. Because this sort of recuperative function is a crucial feature of pains, we need something more than stop imperatives.

Further, what you should stop, and how you should stop it, is context-sensitive. With a cut, I should stop things from touching the cut. With a sprained ankle, I should stop from walking (the *touching* part is irrelevant). Simple contents like "Stop!" cannot make the distinction. Adding information about *how* you should stop or *what* you should stop might help—but then one is most of the way toward protection imperatives.

5.4.2 Proscription Imperatives

A second proposal, and a natural variation on the stop-imperative account, is the idea that imperatives proscribe against motion. I used to defend such an account.[4] Rather than saying "stop!", pains say "don't!"; the pain of a broken ankle says, "Don't put weight on your ankle!" Proscription imperatives can

4 Most explicitly in Klein (2007). Note that in that piece, I also ran together proscription imperatives and stop imperatives (e.g., 520).

handle any case that stop imperatives can handle (for stopping is just a particularly straightforward sort of proscription). But proscription also makes pains conditional standing imperatives, and so can deal with pains of recuperation and the like.

I still find proscription imperatives attractive. Protection will often involve forswearing motion, so in many cases, both kinds of imperatives will end up being satisfied in exactly the same circumstances. However, I think two important kinds of cases weigh against proscription imperatives and in favor of protection imperatives.[5]

The first are cases where pains command passive protection. An attack of gout causes an extremely painful inflammation of the large toe joint. The pain certainly proscribes against *using* the toe, but it also makes the toe extremely sensitive. That is, it motivates the sufferer to *keep things from contacting it.* That seems to outstrip prohibitions on motion: one might be entirely inert, yet still be motivated to keep things from happening *to* your toe because of your pain.

Indeed, for some kinds of recuperation, this sort of protection seems to be more important than straightforwardly limiting motion. Healing a burn, for example, requires that you keep things from contacting the burned skin. That is easy to capture with protection imperatives. It is complicated to handle with proscriptions against motion. For the motions to be proscribed will necessarily be context-sensitive: it might be okay to move your arm if you're in an empty room but bad if you're in a small closet. The pain itself does not seem to change with these changes in context, however. Protection imperatives give a simpler, more natural treatment.

Second, some pains also seem to motivate active protective motions. Being hit in the testicles, for example, provokes a characteristic protective response. Throwing your hands over your groin is surely a good defensive action. Indeed, there are regions of macaque motor cortex that produce stereotyped defensive motions when stimulated (Graziano et al., 2002b), suggesting that the brain generally cares about these sorts of defensive reactions. Defensive reactions seem like straightforward, positive responses, however—it's difficult to interpret them as any kind of proscription on action.

5 Numerous people have pressed me over the years on various kinds of counterexamples to proscription imperatives. For problems in print, see especially Tumulty (2009) and Martínez (2010).

Active withdrawal motions are also protective. Some withdrawal is mediated by pure reflexes. But situations that would produce reflexive withdrawal in the absence of conscious inhibition produce a positive motivation to withdraw *because* they produce pain. I can keep my hand on a radiator for a while, suppressing the reflex to pull back. Eventually the pain will get intense, and I'll have to pull away. This seems like a new, positive action rather than a proscription. Insofar as pains promote active withdrawal motions, then, protection seems more plausible than proscription.

Proscription imperatives are flexible, and it may be technically possible to cast them in a way that avoids the above problems. The first issue can be handled by conditional proscriptions: that is, by imperatives that proscribe conditional on the presence of certain events. The second can be handled by either denying that the positive motions are part of the content of pain proper or crafting negative imperatives such that all motions aside from withdrawal are proscribed.[6] These solutions, however, add considerable complexity to an imperative account. So we should, on balance, prefer protection imperatives because they are simpler and more elegant.

Because I was the main defender of proscription imperatives, I should mention why I gave them up. I was attracted to the idea of proscription because I thought one needed negative imperatives to deal with morphine pain. I think now that there is a more satisfying solution to that problem, to be presented in chapter 11. That solution is compatible with pains being positive imperatives, so I think the theoretical attraction of motion proscription is now far less compelling.

5.4.3 Removal Imperatives

Manolo Martínez (2010) argues for a different imperative content. He claims that pains have a content like:

> Don't have this bodily disturbance!

where "bodily disturbance" refers to some more specific problem. Call these *removal* imperatives.

6 So, for example, one might argue that "The pain of touching a hot stove is a very strong imperative against any action that would keep your hand in contact with the stove (including the trivial action of keeping it exactly where it is). Some sort of withdrawal is thus effectively demanded. However, it is demanded only indirectly, via a command against doing anything but withdrawal" (Klein, 2007, 521).

Like the other accounts, Martínez's account dovetails with protection imperatives in a number of obvious cases. With some extension, it should also be able to handle both passive and active protection cases. However, I think there are several problems with removal imperatives. Just what the problem is depends on how we conceptualize the "disturbance" that is to be removed. Either plausible reading is hard to maintain. First, "disturbance" might refer to some other component of the pain (this move is obviously only available to impure, hybrid accounts). So pain is a state that tells you to remove that state. So, for example, pain might represent damage and tell you to remove the state of representing damage. As I argued in chapter 3, this does not fit well with what I've identified as the biological role of pain. On the one hand, acting in ways that will help you heal won't necessarily remove the pain, except in the long run—my sprained ankle will continue to hurt even as I favor it. On the other hand, removal imperatives would encourage positively maladaptive courses of action. I could probably remove my pain by chopping off my foot; no ankle, no pain in my ankle (if I'm lucky). Less drastically, taking strong painkillers will remove any sensation of pain, but doing so would make me more likely to reinjure myself.

Second, "disturbance" might pick out the actual cause of the pain. Suppose you have a broken ankle. On the one hand, the imperative might command to remove the disturbance *directly*. But that doesn't seem like it makes very much sense: there is nothing you can do *to* a broken ankle to remove the disturbance (in contrast with, say, removing your hand from a flame). On the other hand, the imperative might command you to remove the disturbance *indirectly*: that is, to take any course of action that will eliminate the disturbance. That might include protecting the ankle. It might also include going to a doctor, fashioning a splint, and so on. That seems like far too broad a content to attribute to a simple bodily sensation such as pain, however. Pain causes us to go to the doctor because the fact that we are in pain combines with our knowledge about the typical causes of pain. While that motivation is partly caused by pain, it doesn't seem like a candidate for the intrinsic motivational force of pain—you need to know a lot, and deliberate, before it kicks in. The advantage of protection imperatives over removal imperatives, then, is that protection imperatives command a straightforward, relatively unified type of action.

Third, "disturbance" might pick out *the pain*. That is, what should be removed is the sensation constituted by the removal imperative. I suspect this is closest to Martínez's own account, and I have considerable sympathy for it.

However, I think this confuses pain and suffering. Suffering, recall, gives motivation to remove the sensation. I doubt that sensation removal can account for the *primary* motivational force of pain. For, again, there are many biologically inappropriate ways to remove the sensation. Better, then, to hold fast to the distinction between pain and suffering and cash out the imperative content in some action-centered, rather than sensation-centered, way.

So much for the case against removal imperatives. Here, I think the case is far less obvious than against the other forms of imperativism. For Martínez might well appeal to the *natural* function of pain to handle the sorts of problematic cases I raise for the second reading or deny the distinction between pain and suffering that I use to attack the third account. The disagreement, then, largely comes down to whether we want pain to be constituted by imperatives oriented toward *actions* (as on my account) or toward something else (as on the various readings of removal imperatives). As ever, I think the continuity of pain with the homeostatic sensations argues for an action-oriented story, although it may well be open for the removal-imperativist to simply tell a similar story about the homeostatic sensations.

Perhaps, however, there is a way to split the difference that preserves the intuitive parts of both theories. In chapter 14, I'll argue that Martínez has given a plausible theory of *suffering*. A theory of suffering *should* be sensation-centered, and an imperative relative to a sensation might well fit that gap. For pain, however, we have good reason to prefer the more specific action-centered imperative.

5.5 Conclusion

I have argued that pains are imperatives to protect a part of your body. As imperatives, they convey a command with a content. That content was identified as a set of satisfaction worlds and a ranking function over those worlds. Particular pains convey a certain kind of command: namely, to protect a certain body part, in a certain way, with a certain intensity.

Protection imperatives have further important features. Instances of schema **PS** should be read as conditional standing imperatives. They remain in force until explicitly cancelled. Protection can also come in a variety of different forms. I distinguished between active protection, which requires you to do something, and passive protection, which requires you to *refrain* from doing something. Most pains command a mix of active and passive protection.

Subsequent chapters will show how to explicate these notions within the basic formal framework that I've outlined in this chapter. First, I'll complete the story about imperatives by discussing imperative legitimacy, the attitude of acceptance, and the ways in which imperatives motivate.

6 Motivation and Reasons

6.1 How Do Pains Motivate?

In chapter 4, I distinguished between the primary and secondary motivational forces of pain. Primary motivational force is that which derives from the imperative nature of pains. Secondary motivation is that which is directed toward pain but distinct from it, including all states that constitute a contingent reaction *to* pains. Secondary motivation is a diverse category. It includes, I argued, *suffering* or *hurt*. Motivation because of suffering was separable from primary motivation. That primary sense of motivation is the biologically prior one, because it is what motivates you to protect parts of your body and so keep your body intact.

This chapter will spell out how primary motivation works. My ultimate goal is to address three different challenges.

First, one might worry that imperativism isn't fully intentionalist.[1] Imperative content, the objection goes, can't be intrinsically motivating. Just look at ordinary imperatives. Suppose I tell you to wash my car. You might do it. You might tell me to get lost. In one case you're motivated, and in the other you're not. That doesn't seem to depend on the *content* of my command, however: that remains the same whether or not you obey. Because motivational force is meant to be a phenomenally salient feature of pains, and because pure imperativists like me don't have any *other* content to appeal to, this looks like a problem.

Second, one might worry that imperativism mischaracterizes the sense in which pains motivate us (I owe the objection to David Bain, about whom more shortly). Imperatives just push you around (that's why most people don't like being commanded). But pain does more than push us to act: it gives us a *reason* to act. It is not obvious how imperatives might ground reasons, however. Again, the problem is most severe for the pure imperativist (like me) who distinguishes pain and suffering: I have only a bare command to work with, rather than evaluative qualities such as *feeling bad*.

Third, one might worry that pains don't always motivate. Certain pathological conditions—morphine pain, pain asymbolia, and so on—present us with patients who claim to feel pain and yet are completely unmoved by it. The

1 I originally owe this objection to Todd Ganson; it's expressed, albeit elliptically, in footnote 8 of Ganson and Bronner (2013). Note that Ganson and Bronner assume that imperatives and indicatives have a core truth-evaluable content and so differ in their illocutionary force rather than in their content. I have argued that imperatives differ from indicatives in their content, and that neither should be considered more basic than the other.

problem is not, note, merely that their pain doesn't *hurt*. I've already claimed that suffering is a contingent adjunct of pain even in nonpathological cases. Rather, such patients don't appear to be motivated to *protect* the pained part. If true, that is a counterexample to pure imperativism.

This chapter will spell out the sense in which pains motivate and then come back around to the first two objections to explain how they are answered. The third objection won't be addressed until chapters 11 and 12, where I'll tackle pathological cases. The framework set up in this chapter will, however, provide the groundwork for that later discussion.

6.2 Motivation and Authority

Not all imperatives motivate. Some imperatives—examples in a grammar textbook, say—don't motivate anyone at all. Nor are we motivated by the commands of arbitrary strangers. So we need a story about why and how some commands motivate. That general story will also shed light on imperative sensations in particular.

Here is a quick sketch of the view. Commands that motivate are those that are issued by a *source* that we *accept* as having the *authority* to direct our actions, and so whose commands give us certain *reasons to act*. I will explicate each of the italicized terms shortly.

I take it, however, that the general phenomenon is familiar enough from ordinary-language imperatives. We are motivated by authoritative commands. The traffic cop says, "Stop!" and I stop. I stop because I accept him as an authority. That means I treat him as having the right to issue such commands. His authority means his command also gives me a *reason* to stop: authoritative commands give me reason to act so as to meet the satisfaction conditions that constitute the command. Further, I gain that new reason more or less directly. I don't need to deliberate about whether it is a good thing to stop or why I am stopping: the policeman says "stop!", and I thereby gain a reason to stop.

Not all commands move us in such an authoritative way. Sometimes I follow a command under threat, because of my agreeable disposition, or because I've independently deliberated and come to the conclusion that acting in thus-and-such a way would be good. Although each of these might be ways of *satisfying* a command, they don't count as *obeying* a command. Accepting a command means acting *because* the command has been issued by an authority, and so we treat it as reason-giving. In the basic case, authoritative commands simply give us reasons for action with no further deliberation required. To treat a source as an authority, then, is to be moved by its commands for no other reason than that the source has issued them.

Political philosophers have developed a rich literature on the links among command, authority, and practical reason. I will draw on this literature to flesh out the motivational story. By analogy, the body is a bit like the state. The state issues imperatives in the form of laws and other directives. We are like citizens who accept the authority of the state, and so we are moved by what it commands. We are moved even if those commands conflict with what we would otherwise do and even if they conflict with what we would *want* to do.

My project is not quite the same as the typical one in political philosophy. It is worth noting a few disanalogies. The tradition I'll draw on cares about *legitimate* authority: that is, about what gives a state the right to command its citizens. That is primarily a project about the legitimacy conditions for classes of imperatives.

The current project has a different target. I am concerned with the *attitude* of acceptance one takes toward the commands of an authority. Legitimacy and acceptance can come apart. I can accept illegitimate commands and be motivated by them. I can mistakenly reject legitimate commands and not act on them. Facts about legitimacy are partly extrinsic to facts about the addressee: I may not like a law and may not accept it, but that does not keep the law from binding me.

Of course, the two concepts are not wholly independent. I will assume that accepting a command just is to *treat* it as coming from a legitimate source, whether or not it actually does.[2] From the other direction, legitimate authorities issue commands to affect the behavior of their citizens—which presupposes that their citizens will, on the whole, accept them as legitimate.

I am also concerned with authority in a much weaker sense than most political philosophers are concerned with. Legitimate authorities plausibly derive their standing from the fact that they are constituted by their citizens and make commands on their behalf. Thus, citizens have reason to believe that accepting commands will make them better off overall, following such commands is rationally obligatory, or something along those lines. The legitimacy conditions for this sense of authority thus depend on complex facts about institutions and their relationship to the citizens they serve.

2 In this I am indebted to Raz (1986, 1999).

The body is not quite an authority in this sense. The body is more like a dim-witted, paranoid king: well-intentioned but error-prone and obsessed with self-preservation, and whose commands coincide only roughly with the flourishing of his subjects. There is surely a deep sense in which the commands of such a king are less legitimate than those issued by a constitutional democracy. There is equally a sense, however, in which his subjects accept his authority and are moved by his commands all the same. This sense of authority—what I'll call *minimal practical authority*—is weaker and so more broadly instantiated. It is the sense in which we treat coaches as having the authority to command us to run laps or of grocery clerks to get us to switch checkout lanes. Minimal practical authority nevertheless bears several deep similarities to the more full-blooded sense of legitimate authority. So theories about the latter give useful tools for understanding what is going on when the body commands.

The same structure applies to pains. I have claimed that imperative sensations are commands issued by the body. We accept the body as a minimal practical authority. Its commands thus give us both a first-order reason to act and a second-order reason to continue using those first-order reasons in our deliberations. That is the primary way in which bodily imperatives motivate us. I'll say something about each of these notions.

6.3 The Source of Pains

When the policeman tells me to move along, he is the *source* of a command. All sorts of things can be the source of commands: individuals, groups, states, laws, Pure Reason, and inanimate objects such as signs. I doubt that any intrinsic property unifies this class. All it takes to be a source is to have the ability to express imperatives and the potential to be treated *as* a source by the addressees of those imperatives.

Pains motivate us. So I must take them to have a source. I have suggested *inter alia* that the source is *the body*. It is time to defend that. I'll proceed via exhaustion of possibilities.

For starters, it doesn't seem like pains are issued by anything outside of me. I can feel pain in absence of any external stimulus. My ankle hurts now, although nothing in particular is happening to it. I feel some mysterious pains that I don't take to have an external cause at all. So the viable possibilities seem to be restricted to those that are (in some broad sense) within the skin.

Nor does it seem as if *I* issue imperatives to myself. More precisely, pains are not personal-level self-commands. If so, they would be either voluntary or involuntary. If I voluntarily issued imperatives to myself, then I could presumably stop doing so when imperatives were inconvenient. Pain is not so easy to overcome. Pains come and go of their own accord, and they persist even when I would like them not to. So they aren't like self-commands. Nor do they feel like a kind of involuntary self-exhortation (say, as compulsions do). The perceptual nature of bodily imperatives makes them feel much more akin to something done to us, rather than something we do to ourselves.

That leaves facts about the body. As pains concern the body and promote various actions taken on its behalf, this isn't terribly surprising. But perhaps something short of the body issues commands. Perhaps imperative sensations are issued by *parts* of the body. This would require some care. For as the issuer of the command isn't represented in the content of the command (at least *qua* issuer), merely having a body part as the issuer is not sufficient to ensure that the located body part is part of the content of the imperative. Because the game is to get everything in the content, something designating the issuer must also be part of the content. The most straightforward way I can think to do this is with indexical content. That is, pains might command something like, "Protect *me*!", with the indexical element indicating the issuer. If my finger hurts, then it is my finger alone that I take as commanding me to protect it.

I have a few arguments against treating body parts as issuers. First, pains can concern parts of the body that I don't take myself to have. If I am an amputee, a pain in my phantom hand can't be felt as issued by my hand (I don't think I have a hand). Second, the authority of body *parts* is questionable. Perhaps I have no particular reason to care about my hand and might think that if it offends me, it's better to cut it off and cast it from me. That is still hard to do. However, without a *body*, I can't do anything at all. Formally speaking, this makes for an asymmetry: the authority of body parts is easily treated as derivative on the authority of the body, whereas the authority of the body is difficult to treat as the conjunction of a bunch of disjoint authorities. Third and finally, if the actual issuer and the issuer-as-embodied-in-the-content are always identical, then the former is explanatorily inert when it comes to felt location. It's the body part represented in the command that's really the important part, and the issuer is (even on the body part view) only there to get that contentful part correct.

That leads naturally to the final option: the body is the felt source of imperative sensations. Commands are issued by the body, and they order protecting some particular body part. Again, I mean this as a claim about that which we *take* the source of pains to be. The actual mechanism of generating sensations is some subpersonal process that is not disclosed in the phenomenology. The important fact will be that the body acts as the source, and thus as the issuing authority, for pain sensations.

6.4 Practical Authority

The body issues imperatives. These motivate us by giving us a reason to act. They do so because we accept the body as an *authority*, and authoritative commands give us reasons for action.

"Authority" is used in both a practical and an epistemic sense. It is initially tempting to suppose that the body is an epistemic authority: that is, a source that we trust because we think it has better access to information than we do. David Bain (2011) interprets imperativism in this way.

The body probably can't be treated as an epistemic authority, however. Compare: I am normally motivated by my doctor's edicts, but if I think he is badly mistaken or lacks crucial information, then his commands cease to motivate me. Epistemic authority is undermined if we think we happen to be in a better epistemic position than our source.

Bodily imperatives aren't like that. I can often be quite certain that there's nothing wrong with me, or at least nothing that would be solved by satisfying a bodily command. My fever gives me chills. I know that I am sick, not cold. Nevertheless, my chills still motivate me to curl up under the blanket. Of course, in chronic pain, I may know full well that there's no reason why I should protect my body. The pain nevertheless motivates me to do so, needlessly. So we do not take our bodies to be epistemic authorities.

A practical authority, by contrast, is one whose commands directly guide our actions. Unlike an epistemic authority, we need not believe that a practical authority is actually in a better epistemic position than us to be motivated by its commands (although a superior epistemic position might be the reason why we accept a source *as* an authority). I'll begin by sketching a standard story about full-fledged practical authority. In the next section, I'll weaken that account to give the minimal sense of practical authority that our bodies possess over us.

Hobbes (1651), in speaking of authority, distinguishes command from mere counsel by noting that, "Command is when a man saith do this or do not do this yet without expecting any other reason than the will of him that saith it"

(Ch. XXV). The mark of practical authority is thus that the command alone gives sufficient reason for action.

Suppose that I accept the authority of my department chair, and he tells me to fill out a form. When he does so, I gain a reason to fill out the form. In general, accepting someone as a practical authority means that I take their commands as giving me reasons to act. Commands from a practical authority give me reasons in a different way than my own desires and commitments do, however. The reasons that issue from commands needn't link up, at least directly, with *any* of my antecedent desires or goals. I don't like filling out forms. I would avoid it altogether if I could. Nevertheless, my department chair's command gives me some reason to fill out the form. Accepting his authority is sufficient to give me a reason to act, even in the absence of other reasons.

This points to a first feature of practical authorities, what Hart (1982) calls *content-independence*:

> Content-independence of commands lies in the fact that a commander may issue many direct commands to the same or to different people and the actions commanded may have nothing in common, yet in the case of all of them the commander intends his expressions of intention to be taken as a reason for doing them. It is therefore intended to function as a reason independently of the nature or character of the actions to be done. (254)

That is why my chair's command gives me reason to fill out the form, without further deliberation on my part. It is the fact that I accept the chair as an authority that gives rise to my motivation to do what is commanded—not the details of what is commanded.

Of course, imperatives might motivate in other ways. Again: I might reason that my department chair will be angry if I don't fill out the form, and that I have a desire not to make him angry, and so be thus motivated. Or I might enjoy forms, and so leap at the chance. If one of those is my sole reason for being motivated, however, then I have not really accepted his command as authoritative. For note that I might reason similarly about the commands of someone whom I take to have no authority over me whatsoever. What distinguishes the commands of someone who I take as an *authority* is this direct, content-independent linkage between command and action.

Second, practical authorities give what Raz (1975) calls *preemptive* or *exclusionary* reasons for action. My chair's command doesn't simply give me a reason for action. It also guides subsequent deliberation about that reason. Antecedent to his command, I may have thought that there were good reasons against filling out the form—that it was redundant, that no one would notice its absence, and so on. In accepting my chair's authority, I am thereby excluded from considering those reasons in my future deliberation. Raz (1986) argues one reason why I am so excluded is that my chair should have considered those reasons in his own deliberations. If I were to reconsider them in my own deliberations, then I would be guilty of double-counting (Raz, 1986). Note that this needn't actually be true: my chair might have made a mistake. The attitude of acceptance, however, involves taking authoritative commands as exclusionary reasons. An authoritative command thus cuts off debate, as it were, on the particular merits of the case.

Following Raz, exclusionary reasons are a species of *second-order reasons*. A second-order reason, says Raz (1975), "is any reason to act for a reason or to refrain from acting for a reason" (487). Second-order reasons affect how we weigh first-order reasons. If I know my judgments about distance are unreliable, then I have a second-order reason to downplay any first-order reasons that depend on judgments of distance. If I believe that I am subject to implicit biases, then I have a second-order reason to carefully examine my first-order choices when I am on a hiring committee.

Exclusionary reasons shape our deliberation by giving us reasons *not* to include certain reasons in our practical deliberations. Authoritative commands are one source of exclusionary reasons, but they are not the only one. If I promise to pick you up after work, then I give myself a first-order reason to pick you up. My promise also gives me a second-order reason that excludes reasons against acting as I promised. This is what distinguishes promises from mere predictions: by promising, I bind my future self to act even if it becomes inconvenient. Thus, my promise gives me a second-order reason that should cut off internal debate on acting as I promised.

Authoritative commands work similarly. Authorities provide us with two reasons when they command: a first-order reason to act so as to satisfy their command, and a second-order reason that excludes certain other considerations when we deliberate on that first-order reason.

Command is thus what Hart (1982) calls a "preemptory" form of address:

> ... the expression of a commander's will that an act be done is intended to preclude or cut off any independent deliberation by the hearer of the merits pro and con of doing the act. The commander's expression of will therefore is not intended to function within the hearer's deliberations as a reason for doing the act, not even as the strongest or dominant

> reason, for that would presuppose that independent deliberation was to go on, whereas the commander intends to cut off or exclude it. (253)

Note that this is a slightly more subtle account of what it takes to accept authority than it might seem. It is not that accepting an authority requires giving up on deliberation altogether. Indeed, in many cases, one will still need to deliberate about which of many distinct commands one should satisfy. Rather, acceptance precludes a certain *kind* of deliberation, deliberation about whether to do something on its merits. Acceptance does not cut this off directly; rather, the presence of second-order reasons starves any first-order deliberation of the alternative reasons it needs to work. If your command excludes all other reasons pro and con, and if I take your command as giving me a content-independent reason to act as you say, then deliberation is simply *irrelevant*: there is nothing left on which to deliberate.

We do not accept all commands. Accepting a command requires us to have background commitments that make it reasonable for us to accept a source as a source. I might think that my interests will be best served if I bind myself to the mandates of a legitimate state, for example (Raz, 1986). Other reasons are possible.

These background commitments only serve to set up the attitude of acceptance, however. They don't (or don't normally) enter into our deliberations when we accept a command as authoritative. I listen to my department chair because I want to keep my job, I want the department to succeed, and I want everyone to get along. Although those desires ultimately explain the reason why I accept him as an authority, they don't enter into the proximate explanation of my action in particular cases. It is sufficient that I *do* accept him as an authority, and that he has commanded something, and I am thereby motivated to do that thing.

6.5 The Body as a Minimal Practical Authority

The preceding section gave a sense of what one might call *full-blooded* practical authority. It is clear that we do *not* take the body as an authority in this full-blooded sense. My hunger motivates me, but I don't take it as giving me exclusionary reasons for eating. That is, my hunger doesn't keep me from debating the merits of eating—I take my hunger as only one reason to act among many. Perhaps I am dieting, in which case my diet-related reasons *not* to eat might successfully keep me from acting. Bodily imperatives, in general, don't cut off debate completely.

There is another, weaker sense in which we can treat someone as an authority, however. Suppose the grocery clerk tells me to switch to a different lane. I accept him as an authority in some sense: his command gives me a reason to act, it does so directly, and I accept him as a content-independent source of commands (albeit over a restricted domain). I might have other first-order reasons that I take to override the reasons he gives me—I know the clerk in another lane is faster or that I can only buy bus tickets in my lane. Nevertheless, there's still a sense in which I take the clerk's commands to have a kind of authority: I won't simply *ignore* what he says. His commands must be taken into account; if I don't have a reason for acting otherwise, I'll choose the lane he instructed. I will do so without asking, or being offered, further reasons why.

Call this weaker notion *minimal* practical authority. The commands of a minimal practical authority, I suggest, also give both first- and second-order reasons. The second-order reasons they give are reasons to continue treating the first-order reasons as reasons in our deliberations. Those first-order reasons might then be overridden, but they cannot be simply *ignored*. The second-order reasons of minimal practical authorities are *mandatory* rather than *exclusionary*: they are reasons to continue taking some reasons as reasons in your deliberations, regardless of what other reasons you have.

Minimal practical authority is a weaker notion than full-blooded practical authority. The attitude of acceptance is similar. Both involve taking the authority as a source of content-independent reasons. Absent other considerations, both will motivate by command and solely because the command has been given. The difference lies only in the second-order reasons to which their commands give rise.

We accept our bodies, I suggest, as minimal practical authorities. Our body commands us to protect a certain part. Because we accept our body as a practical authority, that command gives us reason to act—regardless of what else we'd want to do and regardless of what else we know. The reasons that pains give are ones that we continue taking seriously even when we want to do otherwise and even if we know that the body has made some sort of mistake. So they give us both a first-order reason to act so as to satisfy the command and a second-order reason to make that first-order reason a mandatory part of our deliberations. My desire for a pleasant walk might give rise to reasons strong enough to override my pain. But I can't cease taking my pain as a reason to stop: my other reasons must be weighed against it. Further, pain will win out absent some other, stronger reasons that override it.

The motivational structure of bodily imperatives is thus twofold: both a command that provides a first-order reason for action and a mandatory second-order reason that shapes subsequent deliberation. Further, these reasons arise in a direct, nondeliberative way. We gain them merely because we accept the body as an authority. We aren't presented with new *information* about which actions are good, and we needn't think about why the body might be commanding—we are commanded and so accept that command as sufficient to give a mandatory reason for action.

Of course, we accept our bodies as authorities for good reason. Our body is important to us. We care about it. We are in bad shape if it doesn't work. We cease to exist when it does. If we regarded the body as something less than a minimal practical authority, then we would still have reasons to act in ways that promoted bodily integrity. But we would also be free to accept or ignore those reasons as we please. If we did that, then we'd probably screw it up. The demands of the body are often inconvenient. It would be too tempting to put them aside if we could. Overall, we're better off treating our body as authoritative, and thus ceding the right to deliberate on *whether* to listen to its edicts.

So there are many good reasons for accepting the authority of the body. Note, by the way, that "acceptance" here shouldn't be construed as something like *voluntary* acceptance. It's not as if we first deliberate and then decide, all things being equal, that our body is worth listening to. There are powerful evolutionary reasons why the attitude of acceptance should be innate and difficult to overcome.

Treating the body as an authority, however, means that these reasons don't enter into our deliberations on particular actions. That is, when our body commands us to eat, our motivation is explained solely by the command. Acceptance is a background commitment, and the reasons for acceptance do not enter directly into our deliberations. That is good, in most cases. Treating the body as a practical authority solves a commitment problem, as authority often does (Raz, 1986). By accepting the body as an authority, we bind ourselves to at least *considering* the actions that will keep us alive, and we do so in proportion to their urgency.

Treating the body as a practical authority comes at a cost, however. For we are forced to continue doing so even when it commands poorly or when its commands might be positively deleterious. Even in those cases, bodily imperatives continue to motivate, however. The content-independence of authoritative commands binds us equally in good and bad cases.

6.6 Two Challenges Revisited

6.6.1 Motivation and Intentionalism

I noted that imperativism seemed potentially at odds with intentionalism, at least when it came to motivation. We can flesh that argument out a bit more now. Imperative sensations have their content intrinsically (that's the imperativism). That content determines how pain feels (that's the intentionalism). Pain motivates in virtue of how it feels (a trivial consequence, in accord with common sense). But—here is the objection—imperatives can't motivate intrinsically. For imperatives only motivate if you *accept* them. That's an extrinsic fact about the imperative. So imperativism can't explain motivation after all.

Two issues should be distinguished here. The first is whether, given the story I've told, pains count as intrinsically motivating. That's a tricky question and will get hashed out in the course of discussing pain asymbolia. The second is whether, in ordinary cases, the imperative content of a sensation is sufficient to explain why someone is motivated in a particular way, on a particular occasion. That seems to be a much more straightforward issue.

The picture is not that each imperative must be evaluated and either accepted or rejected. If that were the case, then imperatives would clearly be only one half of a jointly necessary causal set, and we'd speak wrongly if we focused attention just on them. Rather, the picture is that we have a standing *attitude* of acceptance toward bodily commands. That attitude is always present, whether or not any commands are issued. It only depends on the fact that we accept the body as a minimal practical authority. We are thus always disposed to accept commands from our body, even when none are being issued, and regardless of what the content of a particular command might be.

Acceptance is therefore part of the general background context against which sensations cause actions. It is, if you like, comparable to the oxygen that makes it possible for a struck match to burst into flame. Except in extraordinary cases, it is a constant. That means that when we give motivational explanations, we keep acceptance fixed; it is part of what Menzies (2007) calls the "default causal model" when thinking about sensory motivation. Imperatives, by contrast, come and go: they are the actual difference makers for particular bouts of motivation (Waters, 2007). Against the background of acceptance, the presence of an imperative is sufficient to motivate. Further, the details of the imperative explain how one is motivated. Nothing further needs to be cited. That is the straightforward sense in which the imperative content of a sensation is sufficient to explain why someone is motivated in a particular way, on a particular occasion.

It seems to me that *any* intentionalist model must similarly appeal to a causal background in this way. That's partly because a relatively complex causal background will actually be necessary to ensure that content-bearing states are possible at all. But it's also partly because causal stories just work this way: outside of the simplest toy examples, *all* causal claims occur against a default causal model. We give explanations, crudely speaking, because we want to know which parts of the world can be manipulated to affect which other parts (Woodward, 2003). We often do that by finding out which parts actually vary and which parts tend to stay fixed. Imperative content is the part of the story that varies quite a bit, in both its presence and manner. That's why it's appropriate to cite it as explanatory and to keep tacit about background facts like acceptance.

Further, indicative contents *also* need some background causal structure in place before representational sensations can have their distinctive causal effects. What really matters for intentionalism, I submit, is that the content alone determines the *distinctive* phenomenal character of a particular sensory state. All intentionalists can admit that various sorts of steady background states are needed before phenomenal character is possible at all.

6.6.2 Motivation and Reasons

In the course of presenting several arguments against imperativism, David Bain (2011) notes two ways in which imperatives don't seem to offer reasons. First, pains give reasons to act by being *unpleasant*. He cannot see how a bare imperative can be intrinsically unpleasant. I agree. But, of course, as I noted in chapter 4, I also think that unpleasantness is a distinct and contingent feature in virtue of which pains motivate.

The contrast between my position and Bain's is important, however. Bain implies that pains motivate *only* because they are unpleasant. I do not deny that most pains are unpleasant, nor that they give you reason to act in virtue of being unpleasant. But they also give you an independent reason to act: because they are *pains*. That is quite independent of whether they are also unpleasant.

I think this claim sounds odd only because pains are so often unpleasant, and so we rarely have a chance to be motivated by pain. But note that the parallel claim is trivial when it comes to the other homeostatic sensations. Hunger can be unpleasant, and that unpleasantness gives you a reason to eat and so get rid of your hunger. But hunger gives you a reason to eat quite independently of whether it is actually unpleasant. Mild hunger is not at all unpleasant. Further, suffering and hunger can motivate in orthogonal ways, just as suffering and pain do. My unpleasant hunger might give me reason to smoke a cigarette and so eliminate my pangs. But smoking a cigarette is not eating, nor is it related to eating. It's a trick to get the hunger to go away. That's surely not what *hunger* motivates me to do, however: hunger motivates me to eat. That's why, again, it is important to distinguish the motivations and reason-giving that are constitutive of pain from the sensation-directed motivations and reasons that come from suffering.

Second, Bain argues, pains cannot give reasons in the right way. He raises, and dismisses without argument, the possibility that pains have something like practical authority. Discussing the reason-giving grounds of pain, he says: "One possible answer cites the pain module's authority. Since a module surely has nothing akin to political authority, the idea might be that it nonetheless has epistemic authority" (Bain, 2011, 181). But we reject epistemic authorities if we believe them to be radically misguided in particular cases, while pains still motivate even when we know that we have no reason to follow them (Bain, 2011).

Fair enough. That's one of several reasons why I think imperativists should treat the body as a minimal practical authority, not as an epistemic authority. That said, I suspect that an appeal to minimal practical authority represents a really deep disagreement with theorists such as Bain. You might find it odd that a source can give reasons for an action without representing the action as a *good* thing to do (in some possibly attenuated sense). Perhaps you even think that there's a constitutive relationship between reasons and the good, such that to give the former just is to represent an action as the latter.

I doubt this, but I don't need to deny it. The phenomenon of practical authority shows that the relationship between reasons and goodness can be more indirect. In the case of pain, what's good is survival, and we accept the body as a minimal practical authority precisely to achieve that end. But that's enough of a relation to the good for particular edicts to give you reason to act. Goodness does not need, as it were, to be *represented* in every command.[3] Nor do particular commands need to be obviously traceable back to the good, save through their relationship to a source you take as authoritative. That is why pains can give us reasons to act, even in cases where we know that the particular action they weigh in favor of will do no good.

6.7 Conclusion

I have argued that pains give us reason to act, and they do so because we accept the body as something like a minimal practical authority. The edicts of minimal practical authorities give us mandatory (i.e., non-ignorable) reasons for action. There are good biological reasons, further, for accepting the body as a minimal practical authority.

Whether that authority might ever be overridden—by either choice or damage—is an interesting question. I'll return to it in future chapters. Before then, I'll flesh out the specific content of pain to handle location, quality, and intensity. As part of this task, I'll justify a choice that has been taken for granted in this chapter. For I've assumed, so far without argument, that it is the *body* that we accept as having the authority to issue commands. As part of telling a story about felt location, I'll justify that choice.

3 I am inspired here by Mark Schroeder's (2008) review of Sergio Tennenbaum's *Appearances of the Good*, and in particular the idea that "good" bears the same relationship to desires as "true" does to beliefs. Thanks to Derek Baker for helpful discussion on this point; Baker also has a useful attack on the Guise of the Good thesis in his "Akrasia and Problem of the Unity of Reason" (draft ms.).

7 Location and Quality

7.1 Fleshing out the Content

Imperativism is a theory about what pains have in common. But to be fully general, imperativism must also tell a story about what *differentiates* pains. Although all pains have some quality in common, they also differ along several phenomenal dimensions. Three are important. Pains differ in felt *location*: sometimes your foot hurts, sometimes your back, sometimes your head. Pains differ in felt *quality*: pains can be stabbing, burning, throbbing, aching, or tingling. Pains differ in felt *intensity*: the same burning pain in your hand might be mild one day and intense the next.

As a pure imperativist, I must account for those differences solely by appeal to variation within the content of the imperative that constitutes pains. My goal will be to show that these differentia can be accounted for in terms of schema **PS**.

PS: Keep B from E (with priority P)!

Capturing the variation among pains using only **PS** is nontrivial. It is here that pure imperativism must work harder than many competing accounts. It is easy to account for intensity, say, by appeal to extrinsic functional properties of pains. It is much easier to account for felt location by introducing a second, representational element. Nevertheless, I think that pure imperativism has the resources to do the job. Indeed, I think it is a feather in the cap of imperativism that it can do so much with so little. I will begin by elaborating some general considerations about imperative content and then move on to using the slots in schema **PS** to differentiate pains. Schema **PS** has three variables that will be filled in by particular pains. B corresponds to a body part, E to a passivized gerund phrase designating the sort of protection required, and P the intensity of the pain in question.

The mapping between E and the more natural English rendering of some pain imperatives is not necessarily straightforward, as the latter often has an implied proscriptive aspect. In "Keep your ankle from bearing weight!", E would not be the property of bearing weight but rather the property of *not bearing weight*. E and B jointly determine a set of satisfaction worlds W_i for the imperative. W_i is some subset of worlds in which $E(B)$.[1] The imperative is satisfied just so long as the actual world is a member of W_i.

Finally, P determines the ranking function $\gtrsim$. Together with W_i, this fixes the content of the pain. Intuitively, the more intense a pain, the higher it ranks its own satisfaction relative to other options. A mild pain can be withstood for the sake of walking to the vending machine, whereas an intense pain precludes it. How P determines $\gtrsim$ is more complex, however, and depends on considerations to be fleshed out in the next chapter.

7.2 Location

7.2.1 Location and Protection

Pains have a felt location. We must take care with this locution. As an imperativist, I must say that pains *themselves* are located, if they're located anywhere, somewhere in the head. This is a general feature of intentionalist accounts. Compare: a sensation as a blue thing over yonder is not located over yonder. It's located in your head and *represents* something as being over yonder. Representational accounts of pain have a similar structure: pains are in the head, but they represent some damage or disturbance at a bodily location, and that location corresponds to the felt location of pain. As long as one accepts some form of intentionalism plus a commitment to transparency of content, you'll get bodily location differing from the location of the sensation.

Similarly so with located homeostatic sensations. The story is trickier, however. Because imperatives aren't truth-apt, they can't represent truths about locations. Unlike the representationalist, we can't appeal to represented facts about locations in the world to explain their felt location. The key, I suggest, is that the felt location should correspond to the part of the body that one is commanded to protect. We feel pains in the location *toward which our concern is directed*. This section will flesh out that intuitive idea.

7.2.2 Bodily Locations

I argued in the previous chapter that pains are issued by the body. If so, how is it that pains can be felt at a particular place?

1 I say "subset" because of further constraints on W_i. As I noted previously, the imperatives of pain are indefinite conditional standing imperatives: that would place further restrictions on when and how one satisfies the imperative.

An analogy is useful. A command by a particular government agency comes from the state and is grounded in the authority of the state. We naturally speak of certain commands as stemming from particular parts of the government, however. When the tax office orders you in for an audit, it is, strictly speaking, the state that commands you and on whose authority you are compelled. Note that if the government otherwise collapsed, such that the tax office was the only bureaucracy with an intact apparatus, it would no longer have either the power or authority to order you around. That shows that the authority of the government as a whole is primary and that of the parts derivative.

Nevertheless, it's quite natural to think of a command to appear for an audit as coming from the tax office. Why? Because that's the only part of the government that's really relevant and toward which your attention is directed by the command.

Similarly so with pains (and located imperatives more generally). The located sensations are just those that command action toward a particular region. We take them to be located in the regions with which they are concerned. My ankle has a pain located in it because my action is directed toward my ankle, and we naturally locate commands in the regions with which they are concerned. While your *body* says "protect there!", it feels as if a particular region says, "Protect me!" The two imperatives would have the same content. Further, your concern is directed by the former to a specific region. It's natural to take that region as a synecdoche for the body. That is why pains are felt *as if* they are issued by the locus of concern, although they are more strictly speaking issued by the body as a whole. Including bodily location in the content proper, and linking felt location to that location, means that the felt location of pain will always coincide with the location toward which pain directs your concern.

Not all imperative sensations are located. Mild hunger and thirst, say, aren't felt in any particular location.[2] The present account explains that fact nicely. Hunger and thirst don't direct you toward any particular body part but rather toward actions taken toward the body as a whole. Conversely, imperative sensations such as itches, which command you to scratch a particular location, are also felt as located in particular regions precisely because they direct activity toward that region. Finally, some plausibly imperative sensations such as

2 More severe versions aren't felt as located either but are sometimes confused as such because of accompanying sensations such as dry mouth or stomach cramps. But put those aside; the mild versions are enough to make the point.

fatigue or nausea can be felt throughout the whole body. The actions they direct are ones that involve the whole body (resting, refraining from any activity, or something along those lines), again consonant with the theory. So the story about the location of pains is quite generally applicable to the imperative sensations, located or otherwise.

This implies that pains' location is determined by the term *B*. The difference between a mild burning pain in your foot and a mild burning pain in your hand, then, is that the former commands you to protect your foot, whereas the latter commands you to protect your hand. It is properly speaking your *ankle* or your *arm* that hurts, not a bunch of locations that happen to be collectively in your arm. That's entirely appropriate. It is the broken ankle, considered as a functional unit, that needs to be protected from weight-bearing.

This sheds light on some hard-to-grasp pains. It is a curious fact about pains that some of them are hard to localize. Chronic back pain is often felt only as emanating from the lower back but with no more specific locus. The imperativist account handles this nicely, for bodily locations can be *functional units*, with pains felt at particular locations within that unit only derivatively. Pain in the lower back is felt in the lower back as a *whole* rather than as numerous individual pains at each location in the lower back. That is why close introspection on diffuse pains leads to unreliable intuitions about location.

Of course, there are also cases where pains attach to arbitrary or nonfunctional units: a burned patch of skin, say, may cross several dermatomes, and thalamic pain syndrome covers half the body.[3] Even in these cases, however, I think the pain *creates* a functional unit. Your concern is with *the burned patch*, considered as a temporarily coherent whole. Particular locations within the patch are important, and worthy of your concern, only because they are part of the burned patch.

Thinking of pain location in this way has an interesting and important consequence. Strictly speaking, there is no sense in which pains can be *mislocated*. The imperatives constitutive of pain direct your concern to a particular part of the body, and that part determines felt location. That location might be independent of the location of the cause of the pain, as it is in the case of referred pains. Aydede (2006a) notes that representationalists must claim that referred pains are mislocated: if pain is a representation of a disturbance in a particular place, then it might misrepresent that disturbance. That seems at least a bit

3 Thalamic pain syndrome, an occasional consequence of small strokes to the thalamus, can produce severe pain that is localized to only one half of the body (Ramachandran et al., 2007).

counterintuitive: it would be a little odd if pains were somewhere other than where you felt them.

This is, I suggest, because the felt location of a pain is partly constitutive of it, and so it's hard to see in what sense location could go wrong. Imperativism gives a more straightforward answer. In classic referred pains, the pain is felt at some distance from the site of the disturbance—the pain of a heart attack is often felt in the arm. But a referred pain introduces only a disconnect between the actual source of the sensation and the felt location of the pain. The felt location of a referred pain and the body part that one is commanded to protect still coincide: heart attack pain in the arm still motivates you to protect your arm.

Imperativism preserves this intuition while explaining the sense in which referred pains are maladaptive: referred pains are felt as located where the body directs its concern. But concern is misdirected in this case—it is your heart you should be worried about, not your left arm. Thus, referred pains can be *mistaken* and *maladaptive* without being *mislocated.*

7.2.3 Pain, Body Image, and Body Schema

7.2.3.1 The Distinction

It's entirely possible to have your concern directed toward something that is not actually part of your body, and so to feel pains located where there is no body part at all. Phantom limb pain is a well-known example. The phantom sufferer is commanded to unclench his hand. That command is satisfied in worlds where he has a clenched hand and unclenches it. Because he has no hand, he can't satisfy it. As I will explore further in chapter 9, that makes phantom pains unsatisfiable commands. Concern is directed toward a region that is, in fact, mere empty space. That is why our sufferer can't do anything about it. Nevertheless, pains can be located *where the hand would be* if there were one because they direct concern to that (nonexistent) body part.

Such is the story. However, that complicates my story about bodily locations. Whatever locations are, they can't be unproblematically identified with bits of the actual body.[4] So it's worth thinking a bit more about the sense of "bodily

4 I assume we also want a story that distinguishes bodily locations in part so that people who share bodies might still feel distinct pain. In case you were worried about that sort of thing, perhaps because of Wittgenstein's (1953) claim that it would "also be imaginable for two people to feel pain in the same—not just the corresponding—place. That might be the case with Siamese twins, for instance" (§253).

location" to which the imperativist should appeal. Here I have no firm story to tell, but I have several suggestions.

To begin, the relevant sense of bodily location seems to depend on something like an internal *representation* of the body. The standard story about phantom pains, for example, is that they result from the persistence of the body image in the absence of an actual body part (Ramachandran, 1998; Ramachandran and Blakeslee, 1999; Ramachandran and Hirstein, 1998). This body image determines which body parts we treat as *perceptually* present, even if we know better. Hence, it's possible to have concern directed toward an actually nonexistent body part because the sense of bodily location is inherited from this body image.

Recent scholarship, however, has distinguished at least two different types of bodily representation. Most authors now distinguish between the *body image* and the *body schema* (de Vignemont, 2010; Gallagher, 1986, 2005). Following de Vignemont, I'll treat the body schema as covering all "sensorimotor representations of the body that guide actions." The body image, in contrast, is composed of representations of the body that are not primarily action-oriented (de Vignemont, 2010). Treating these terms as singular is a simplification: both should properly be understood as *categories* that subsume many different bodily representations. What's common to any representations in the body schema category is that they concern the body as the source of our *actions* in the world. The duality of body image and body schema mirrors, in this sense, the duality of the body: our body is both an object in the world that can be acted on and the thing with which we act.

The distinction between the two types of representation is complicated by both the causal interactions between different sorts of maps on both short and long timescales (de Vignemont, 2010; Schwoebel and Coslett, 2005) and the overlapping nature of many cortical representations (Graziano and Aflalo, 2007). Nevertheless, there appears to be decent empirical evidence for the distinction. If so, then we might ask to which family of bodily representations pains should direct their concern.

7.2.3.2 Body Image

"Body image" is a negatively defined category. It includes all representations that are not primarily used for direct action guidance. Representations in this category are sometimes divided into perceptual and semantic (Schwoebel and Coslett, 2005). Perceptual body image representations portray visual, tactile, or other straightforward properties of the physical body *qua* object, whereas semantic representations cover known facts about body parts. Semantic representations don't seem to be relevant for pain, so I'll confine my discussion to perceptual body images.

There is some evidence for the involvement of body image in pain perception. As noted above, the standard story about phantom limbs involves the inappropriate persistence of a perceptual body image after the loss of the affected limb; further, the same process appears to be involved more widely in gender dysphoria, apotemnophilia, and similar problems with body identification (Brang et al., 2008; Ramachandran and McGeoch, 2007). Further, patients with chronic pain syndromes often reveal distortions of body image when (for example) asked to draw the affected part of their body (Lewis et al., 2007; Moseley, 2008).

That said, I find the evidence for involvement of the body image—considered as distinct from the body schema—in pain perception to be relatively weak. Ramachandran does not appear to consistently distinguish body image and body schema in his discussion of phantom limbs. Further, there are direct arguments that congenital phantoms involve the body schema (Gallagher et al., 1998). Distortions of perceptual reference might just as well be a consequence of cortical reorganization, or else of problems with peripheral input, rather than evidence for the importance of the body image in pain.

Further, the body image doesn't seem like the right sort of thing for an imperativist to appeal to. The body image, recall, includes all representations that treat the body as the *target* of actions (including verbal reports) rather than the *effector* of them. On a superficial level, the imperativist might be read as treating bodily locations as targets of protection. I think this is wrong, however. For the primary effect of the imperative of pain—as I've continually emphasized—is to get you to take certain kinds of *action*. Protective actions might have the body as a target (as with pointing; see de Vignemont, 2010), but they are primarily a way of *acting*. Further, being commanded to take protective actions should have a strong effect on the actions that seem salient and appropriate for us to take.

7.2.3.3 Body Schema and Defensive Representations

These considerations suggest that pain should have a relationship to the body *schema* rather than the body image. There is also evidence for the involvement of the body schema in pain. Patients take longer on motor imagery tasks with their pained limb than an unaffected one, for example (Schwoebel et al., 2001). Similarly, patients with back pain are less accurate at motor imagery tasks involving trunk rotation (Bray and Moseley, 2011). Because motor imagery tasks are generally thought to tap into the body schema, this suggests some connection between the protective function of pain and the body schema.

A link between pains and the body schema would be clearest if pains merely limited motion. As I argued in chapter 5, however, the protective function of pain seems to be more complex. Some pains keep us from using a body part, whereas others command us to passively protect a body part. The latter sorts aren't straightforwardly accomodated on a simple body schema view.

One might have pain affecting both the body schema and the body image, but that's tedious and messy. A better possibility, I think, is to posit finer-grained distinctions within the body schema as well. The cortex appears to map and remap everything, after all, so it wouldn't be surprising if body schemas were as multiple and heterogenous as body images. The idea would then be that there's a body schema representation which is primarily concerned with *protective* action: that is, one which maps out parts of our bodies that we should pay special attention to, avoid using, keep from contacting things, and so on. Call this a *defensive representation* of the body: it shows which parts of the body are in need of which sorts of defense. Defensive representations would interact with other body representations (including short-term action schemas), and so would shape and provoke appropriate responses. Pain would naturally fit into such a framework, with bodily locations in **PS** referring to bits of this defensive schema. Of course, without a further story, this proliferation of body schemas might seem uncomfortably *ad hoc*. Here, I've been deeply inspired by Michael Graziano and colleagues' work on the function of the motor and premotor cortex.

Traditional stimulation experiments suggested a roughly somatotopic organization of the motor cortex, along the lines of the somatosensory cortex. Different sites were thought to represent muscles or perhaps muscle groups. Experiments showing somatotopy, however, relied on short bursts of stimulation. Graziano et al. (2002a) used realistically long stimulation (on the order of 500 ms) in the monkey motor cortex. They found that such stimulation actually provoked complex actions rather than mere muscle twitches. These complex actions appeared to be systematically organized within the cortex, with nearby areas provoking similar gestures. Further, different parts of the cortex appeared to provoke different *kinds* of movement. These included distinct regions for grasping, complex manipulation of central space—and defensive movements.[5] Further evidence suggests a complex fronto-parietal network for building and defending a zone of personal space that might be implicated in a variety of defensive actions (Graziano and Cooke, 2006; Kaas et al., 2013).

I don't suggest that the defensive network identified by Graziano is the same as the body schema I've proposed. (Indeed, I'll suggest in later chapters that the insula is probably a more promising site for localizing such representations.) Rather, it provides two distinct sorts of evidence. First, the multiplicity of action representations in the motor cortex is good evidence that the "body schema" does not pick out a single thing but can be subdivided into a variety of action-oriented representations. Second, it shows that at least some of these representations are oriented toward defense of the body and its local space. That really shouldn't be too surprising. A running theme of this book has been that we have extremely good reason to be concerned about the state of our body, and that in turn affects what we can and should do. Graziano et al.'s work focused mainly on actions that defend against things coming into local personal space. My suggestion is that there should be a similar representation of actions necessary to protect the body. It is then this representation of the body to which pain location information refers.

7.3 Quality

7.3.1 The Quality of Pain

Pains differ in quality: they can be throbbing, burning, aching, stabbing, twinging, intermittent, sickening, pinching, gnawing, and so on. The McGill Pain Questionnaire (MPQ), a standard instrument for assessing pain, lists 102 different descriptors for pain organized into three major classes and sixteen subclasses (Melzack, 1983). A good theory should account for these distinct qualities of pain. Saying just what counts as a quality of pain, however, is tricky. To begin, I will do a bit of divide-and-conquer to carve up the explanandum class.

The MPQ is as good a place as any to start, and its divisions are handy for the job. (Note that the descriptors of the MPQ aren't meant to be mutually exclusive. Pains can, and often do, have multiple descriptors drawn from a variety of categories.) Of the MPQ's three large divisions, two won't be relevant. The first, "Evaluative" set of words largely relate to pain intensity, which

5 In Graziano et al. (2002b), see especially figure 4 for an example of defensive postures. See Graziano (2006) for a broad review.

will be covered separately in chapter 8. The second, "Affective" division contains descriptors such as "fearful," "tiring," "sickening," and so on. These don't seem to pick out properties of the pain. Rather, they relate to emotions *evoked by* pain or to a subject's *evaluation of* the pain—that is, to qualities relevant to the secondary motivational force of pain. These facts are important in a broader account of pain experience. They may be particularly useful in medical contexts. But these are qualities of mental states caused *by* pains, not qualities *of* the pains. We may pass over them for present purposes.

7.3.2 Pattern Qualities

That leaves the third, "Sensory" division, which seems to track unique variations in pain. Within the sensory, we must make a final distinction.[6] Some of the sensory descriptors are qualities that a pain can have at any particular instant. Call these the *sensory qualities* proper. Others refer to spatio-temporal patterns that pains exhibit. Call these *pattern qualities*.

Pattern qualities are had by pains in virtue of their variation in intensity and location over time. Pattern qualities are picked out by descriptors such as "flickering" or "shooting." The former picks out a pain with a certain kind of temporal variability: it makes relatively minor changes in amplitude over time at a relatively high frequency. The second picks out a pain that moves location quickly.

The first two subcategories of the MPQ sensory descriptors ("temporal" and "spatial") each appears to pick out pattern qualities.[7] Some of the other sensory descriptors may also pick out pattern qualities: "stabbing" or "drilling" pains might receive a similar treatment to shooting pains, for example, save that the spatial movement is perpendicular to the bodily surface. I think that's not entirely clear—I suspect, for example, that the stabbing pains also have a distinct sensory quality—but it is an empirical question for each descriptor which of the two types of quality it picks out.

Pattern qualities are not part of the content of pain. Nor need they be. We can account for them by postulating certain more basic qualities of pain plus a

6 Originally made, at least to my knowledge, by Michael Tye (1995). In that work, Tye doesn't completely distinguish between pattern qualities and sensory qualities proper, at least insofar as I read him; the distinction is clearer in Tye (2006).

7 To make matters more confusing, the MPQ includes a second set of temporal descriptors such as "brief," "transient," "steady," and so on. These strike me as similar in kind to the temporal sensory descriptors, save that they cover temporal patterns over longer timescales. So I'll assume that they admit of an identical treatment.

suitably introspective agent. The reason that you are aware of a shooting pain is because you (a) have a pain that varies, and (b) are aware of that variation. So being aware of a shooting pain is being aware of a fact *about* a pain, not being aware of a quality of the pain. This might appear to be in tension with my intentionalism. But intentionalism is perfectly compatible with introspective facts about mental states. Every intentionalist thinks that you could be aware that one sensation occurred before another, and that this fact needn't be traceable to the content of either sensation. All you need is two sensations, plus appropriate introspective access. The present account of pattern qualities simply extends that plausible idea.

7.3.3 Sensory Qualities

Not all sensory descriptors pick out pattern qualities. A burning pain, for example, seems qualitatively distinct from an ache. Both are distinct from cramps and stinging pains. This final set of sensory qualities, then, probably can't be reduced to combinations of other qualities. They need separate explanation.

One common suggestion is that these remaining sensory qualities depend on representations of particular types of injury. So, for example, burning pains represent burns, stabbing pains represent stabs, and so on. Obviously I can't appeal to such representations. But such accounts are independently implausible. For one, there are terms (such as "ache") that pick out sensory descriptors but that don't seem to pick out particular types of damage. So it seems that the representationalist view is based here on a particular quirk of English. Second, pain descriptors don't correlate with the type of damage that descriptor would intuitively be taken as describing. Cuts are often described as burning, stabs as sore, fractures as hot, labor as stabbing, and so on (Melzack and Katz, 2006; Melzack et al., 1982). I suspect that we learn words for sensory qualities from the words that others use, not by reference to the type of injury.

The imperativist, by contrast, can account for the remaining sensory descriptors. First, note that sensory qualities are categorial rather than dimensional: that is, there are a bunch of different *types* of qualities, each of which is simply distinct from the other. These categories don't exclude one another, although there may be *de facto* clusterings of categories.

Note too that the *kinds* of protective action that pains command also fall roughly into categories. In chapter 5, I distinguished active and passive protection imperatives and noted that individual pains might vary in the mix of active and passive protection that they demand. Some pains proscribe use, some demand vigilance against touching, some warn against continued muscular action, and so on.

I propose, then, that different sensory qualities correspond to *different kinds of protective action* that one might take. (Here there is a natural affinity with the story about locations and the defensive body schema that I gave in section 7.2.3.3.) The difference between the sharp acute pain when you touch a stove and the later particular dull ache as your burn heals is a difference in what you do to protect the area. When you touch the stove, you protect your hand by removing it as quickly as possible. When your burn is healing, you protect the same patch by treating it gingerly. Different demands lead to different feels.

This means that sensory qualities proper depend on the term E in schema **PS**. E, recall, picked out the sort of action to take with respect to B, and the satisfaction conditions of the imperative were just the worlds in which $E(B)$. The present suggestion is that Es do not vary arbitrarily but instead fall into broad categories. The sensory qualities of pain track those categories.

The imperativist account of quality is ultimately an empirical claim. Distinguishing different types of protective action is a bit tricky. I'm a philosopher, and so I am limited to scouring the extant empirical literature. What there is, however, seems at least prima facie plausible. Both fractures and sprains, for example, are frequently referred to as aching (Melzack et al., 1982, Table IV). Fractures and sprains don't have much physically in common, but they do require the same sort of protective action—that is, avoidance of weight-bearing and use. Cuts, fractures, and bruises are often described as hot or burning, suggesting proscription against contact. Some pains—such as cramping pains—obviously relate to the cessation of muscular activity.

One might also look to the extensive literature on "first" versus "second" pain (Price and Dubner, 1977). Pains after acute injury often come in two waves. The first wave is typically sharp and well localized both spatially and temporally. The second wave is slower to arrive, usually lasts longer, and is felt over a wider region. First and second pains have a distinct phenomenology. As Price et al. (2002) describe them,

> ... investigators observed that first pain was sharp or stinging, well localized, and brief, whereas second pain was diffuse, less well localized, and had qualities of "dullness" "aching" "throbbing" or "burning." (597)

Most work on first versus second pain has focused on the differential transmission times of peripheral Aδ and C-fibers, which explains the temporal gap between the two waves. Yet first and second pain also seem to correspond to different protective demands: first pain corresponds to the need for rapid withdrawal, second pain to the need for post-injury protection. The difference in phenomenal quality between the two, I suggest, reflects precisely this difference in what sort of protection is commanded.

In summary, then, I suggest that insofar as there are categories of sensory qualities of pain, those categories track categories of protective action. A mix of sensory qualities and pattern qualities is, in each case, sufficient to account for the quality of pain.

Of course, the account of quality might admit further elaboration. Our ordinary descriptors pick out broad similarities between pains. But it's entirely possible that specific pains have their own complexities in what they command. The Schmidt et al. (1984) work on the "Schmidt Pain Scale" for insect stings is a notable example.[8] Schmidt not only ranked the pain caused by several dozen different varieties of stinging insects but also provided picturesque descriptors such as "turning a screw into the flesh" and "ripping muscles and tendons." Given the complexities of individual body parts, the activities possible to perform with them, and the insults it's possible to subject them to, it's quite possible that individual pains might have more specific qualities determined by specific demands for protection.

8 See also Starr (1985) for a similar scale. Thanks to Adrian Currie for drawing my attention to Schmidt.

8 Intensity

8.1 Imperative Intensity

Pains differ in *intensity*: they can be mild, severe, or anything in between. Differences in intensity reflect differences in motivational force. Mild pains are easier to override, whereas severe pains almost inevitably force compliance. I argued that intensity corresponds to the term *P* in schema **PS**. But what *P* amounts to has not yet been explained. One could, of course, simply treat *P* as a brute fact—something like a simple number that allows comparisons among pains and (perhaps) between pains and other sensations. Absent some further story, however, this would be unsatisfying and might open imperativism to charges of special pleading. It is not obvious—at least from the outset—that a simple number could be part of the *content* of pain. For all I've said, intensity might be equally well treated as a *functional* property of pains, one that could furthermore vary independently of content. So the pure imperativist really should say something more about intensity.

As ever, the strategy will be to draw an analogy with ordinary language imperatives. Ordinary imperatives can also vary in something like intensity. Further, that variation can and *should* be modeled as part of the content of imperatives. A variety of linguistic phenomena suggest that variations in intensity must be captured as variations in content.

In chapter 5, I suggested that imperative content is more complex than a simple set of satisfaction worlds. At least some imperatives that differ in content differ only in how they *rank* the worlds. Using a framework developed with Manolo Martínez, I'll argue that these ranking functions can be used to ground differences in the intensity of imperatives, and then I will translate that story into one about the intensity of pains. (Credit where credit is due: when I say "we" and "our," I mean Martínez and I. The full account is spelled out in our "Naturalism and Degrees of Pain," forthcoming.)

In telling this story, one of my goals is to keep it as general as possible. In particular, I want to remain neutral on two important issues. First, I want to remain neutral on what sort of thing pain intensity actually *is*, in particular whether it is a true psychometric magnitude. Some pain scales are straightforward ratio scales for pain, and ratio pain scales do seem to have considerable internal validity (Price et al., 1983). This finding suggests that intensity is a cardinal function: all pains are intercomparable and can be represented as a single number on a scale with a definite zero. Some have doubted the validity of ratio scales, however, and suggested that only ordinal pain comparisons are really possible (Hall, 1981). That is, it might be that some pains are more or less intense than others, but that pain intensity can only be compared rather than ordered.

I intend the account that follows to be neutral between the two options. If pain intensity is a true psychometric magnitude, then the ranking function I'll appeal to could be trivially derived from those magnitudes. If intensity can only be compared, however, then it could still be captured by a ranking function. Under the right conditions, we can show how judgments of pain comparisons will correspond to a ratio scale, even though the underlying quality isn't a true magnitude. The latter case is the more complex one, and so the source of the challenge.

A second issue—and one that's not discussed in the literature, at least as far as I can tell—is whether pain intensity admits of ratio comparisons to other motivating states such as hunger, thirst, and so on. My intuition is that the homeostatic sensations generally admit of only ordinal intercomparisons. That is, it makes sense to say, "The pain in my foot is more intense than my hunger," but not things such as, "The pain in my foot is twice as intense as my hunger." The first sentence sounds fine to my ear, the second sounds bizarre. If you have that intuition as well, then you'll probably want intensity represented by something more complex than a mere number, because mere numbers could be compared across modalities (or, at least, you'd have to do some fancy footwork to make them orderable but not ratio-comparable).

Accounting for these two possibilities is what leads to most of the complexity in this chapter. If you're confident that pains are true magnitudes and their intensity is ratio-comparable to other sensations, then you should probably just treat intensity as a simple number. A simple numeric intensity gives rise to a straightforward $\gtrsim$, and that, along with the set of satisfaction worlds W_i derived in the last chapter, completes the full story about pain's content. If you're unsure, or if you want to keep open the possibility that pain might be more complex, then the full framework will be of interest.

8.2 Intensity

8.2.1 Clarifying the View

Pains differ in intensity. Ordinary imperatives also differ in intensity (or *force* or *urgency*). All things being equal, shouting, "Pass the salt now!" conveys a more forceful command than does a calm, "Please pass the salt at your earliest convenience!"

Intensities might be loosely ordered, as in the case of the two commands about the salt. Force may be more precisely individuated, as it is in some formal contexts: triage classifications in emergency rooms, the rubber stamps used by enthusiastic managers to prioritize their edicts, or the priority codes on the Autovon telephone network.

The Autovon network is an instructive example. It was developed during the cold war to provide communication during a nuclear attack. An Autovon keypad had an additional column of four keys that allowed the user to specify the precedence level of the call. Higher precedence calls would, if necessary, kick lower precedence calls off the trunk. This ensured that more important calls could go through. (After a nuclear war, the thinking went, not all calls could be completed.) The highest level, which would guarantee that a call would go through if any could, was restricted to use by the White House. Note, importantly, that the urgencies here relate to the *routing of the calls*, not to the contents of the calls themselves.[1] The White House could theoretically use Autovon to call Pizza Hut. While the call would trump all others, delivery during a nuclear war would probably be hard to guarantee.

So it seems like ordinary imperatives can vary in something like force, and that variation has something to do with how likely or urgently it is that they be satisfied. Pains also vary in intensity. A natural connection: variations in intensity of pains are due to variations in the intensity or force of the imperatives that constitute them. Intensity or force, we suggest, can be cashed out in terms of the ranking function $\succsim$: crudely, more intense, urgent imperatives will, all things considered, rank their own satisfaction higher than less intense ones. For example, PRIORITY calls on the Autovon network rank their completion more highly than ROUTINE calls, and vice-versa.

Before I start, it's worth distinguishing intensities from a few other things that look a lot like intensity but aren't. Failure to do so has led to some objections to imperativism, so it's worth dispatching those.

First, the intensity of a command should be distinguished from the *illocutionary force* of an *utterance* of an imperative. Imperatives can be polite or nasty. They can be phrased as requests, pleas, or straightforward commands. None of these is a variation in intensity per se, although one might *convey* variations in intensity by variations in this illocutionary force.

1 Technical aside: I treat the routing precedence as part of the content of the command to connect to a particular end user. This command occurs together with the phone number, which specifies the end point.

Failure to distinguish the two may lead to the impression that variations in intensity are not part of the content of imperatives. In a recent paper, Cutter and Tye (2011) object that an appeal to degrees amounts to abandoning intentionalism. Regarding the proposal that imperatives vary in intensity, they write that:

On such a proposal, the difference between [two pains] is analogous to the difference between the following two imperative sentences:

- (Please) stop that bodily disturbance.
- Stop that bodily disturbance!!!

They then object that such an account is inconsistent with intentionalism because,

[N]ow the phenomenal character of an experience does not supervene on its content alone; rather it supervenes on its content together with its degree of urgency. (Cutter and Tye, 2011, 104)

I agree with Cutter and Tye that the two sentences above don't differ in content (or at least they need not). But it's not obvious that they differ in their intensity either. The two sentences above vary in *politeness* not in the force of the imperative. More generally, two imperatives may differ in illocutionary force without differing in content. The same command may be ordered, demanded, requested, or politely suggested. Why? Because commands are individuated by their content—on my account, the satisfaction conditions of the command plus a ranking function. Differences in illocutionary force depend partly on the social circumstances under which imperatives are uttered. So the same imperative can be conveyed in varying external circumstances and can carry additional information relevant to those circumstances and irrelevant to the satisfaction conditions of the command it expresses.

Illocutionary force creates the potential for confusion, I suggest, because sometimes intensity can be conveyed by uttering imperatives with a certain variety of illocutionary force. Rude commands tend to suggest high priority, for example, and polite ones low priority. But note that this is entirely compatible with priority being part of the *content* of the command. Compare: I might convey the temperature of a 110-degree day by assuring you that it is "very *bloody* hot." My rudeness conveys the degree to which some external magnitude varies. Yet my representation of different intensities of temperature still supervenes on the content of my representation. *What* I've conveyed about the temperature by my rude claim is (say) identical to what is conveyed by the neutral, "It's more than 100 degrees out." Similarly with imperatives: rudeness is striking, and so we can use rudeness to convey that a particular imperative is urgent. Yet in each case, these extrinsic facts are being *used* to convey an imperative with a certain intensity; they don't constitute the intensity itself.

Second, I will sometimes use "urgency" as interchangeable with "intensity" because in many contexts it sounds more natural. Intensity is not urgency in the sense of *temporal* priority, however. Some imperatives should be satisfied sooner than others. This temporal ordering usually depends, at least in part, on the content of an imperative. However, temporal priority is an *extrinsic* property of imperatives. That urgent bit of dusting becomes less urgent when the house is on fire; the urgent ache in your toe suddenly becomes less so when the bear appears. Of course, temporal priority depends on intensity: all things being equal, a more intense imperative should also be satisfied sooner. But it also depends on what else is going on and which other imperatives you are subject to. So while the temporal priority of an imperative is tightly related to its intensity or urgency, it is not the same thing: what's important is *formal* priority relative to other imperatives.[2] Temporal priority is an extrinsic fact about an imperative, whereas formal priority isn't (or need not be). Just how that works will be demonstrated shortly.

8.2.2 Ordinal Comparisons of Imperatives

Two observations are the key to a comparative theory of imperative priority. First, note that the $\gtrsim$ for an imperative i ranks outcomes over all possible worlds. This means that it will also include worlds in which *other* imperatives are satisfied. Second, note that $\gtrsim$ is not restricted to ranking worlds in W_i (the satisfaction worlds) as higher than worlds not in W_i (the non-satisfaction worlds). That is, it is possible to have an imperative that ranks the satisfaction of some *other* imperative as strictly better than its own satisfaction.

That may seem odd. But note that it is precisely the case with formal structures of imperatives like those discussed in section 8.2.1. An Autovon call with priority IMMEDIATE should be routed to its destination. So all things being equal, its associated $\gtrsim$ will rank worlds where the call is completed higher than ones where it isn't. Further, it should be routed preferably to calls with priority

2 Contrary to what I suggested in Klein (2012). Thanks to Manolo Martínez for talking me out of it and to Elster (2009) for clarifying the relationship between urgency and temporal priority.

ROUTINE. So worlds where it is routed and a ROUTINE call is not are ranked as strictly better by $\gtrsim$ than worlds in which the reverse is the case. Finally, sometimes all things aren't equal. A call with priority IMMEDIATE should be *dropped* in case of conflict with a call of priority FLASH. So there are non-satisfaction worlds—ones in which the call fails in favor of a FLASH call—that will ranked by $\gtrsim$ as strictly better than satisfaction worlds (i.e., where the call goes through at the expense of the FLASH call).

Let's say that an imperative i is satisfied *at the expense of* some imperative k just in case either i is satisfied and k is not, or that i's satisfaction delays k's satisfaction. (Note that in the following examples, k might be a merely hypothetical imperative for the purposes of ranking, not one that an agent actually entertains or is bound by.) Say that an imperative i is *semipreferable* to an imperative j relative to some imperative k just in case either:

1. i ranks all worlds in which it is satisfied at the expense of k as better than worlds in which k is satisfied at the expense of i, and j does not rank all worlds in which it is satisfied at the expense of k as better than worlds in which k is satisfied at the expense of j; or
2. i ranks some worlds in which it is satisfied at the expense of k as not worse than worlds in which k is satisfied at the expense of i, and j ranks all worlds in which it is satisfied at the expense of k as worse than worlds in which k is satisfied at the expense of j.

Given this, we define imperative priority as follows:

> *Priority* An imperative i is more urgent than an imperative j just in case there is some imperative relative to which i is semipreferable to j and no imperative relative to which j is semipreferable to i.

Let's unpack that a bit. The simplest cases—and by far the most common—will be ones in which i ranks its own satisfaction as better than j's, *and j agrees*. How do we know which of two Autovon phone calls is more urgent? Well, a FLASH call i ranks its own connection as better than some ROUTINE call j, and the ROUTINE call j also ranks its completion as less important than the completion of the FLASH call. So i is semipreferable to j. As that structure is consistent across the different Autovon levels, j is never semipreferable to i. Hence, i has a higher priority than j. Note here that when sorting out semipreferability, the third imperative k can be set identical to j.

The third imperative k and the indirect structure it makes possible are necessary for a tricky subset of cases (ones, however, that pains arguably exemplify). Consider the imperative, "Repent your sins, and the sooner the better!" The pastor and the prophet might issue these commands in a way that varies in intensity—the prophet utters it as a matter of gravest importance, whereas the pastor is more understanding of the complexities of modern life. Yet the two imperatives have the same satisfaction conditions W_i. Further, they always rank earlier satisfaction as *ceteris paribus* better than later satisfaction. So it is not obvious how to drive a wedge between the two. Note that this problem appears whenever there are two imperatives with the same W_i: it is not possible to construct worlds in which one but not the other is satisfied, or where one is satisfied earlier than the other, and so it's difficult to see how they might be ranked against one another.

Here, however, one can still construct semipreferability by reference to *other* things one might do. Consider some third imperative k—say, to move your car from the fire lane. The prophet cares not for fire lanes; $i_{prophet}$ ranks repentance higher than satisfying k. The pastor understands the importance of rendering unto Caesar and all that. So i_{pastor} might rank satisfying k now as more important than repentance. (Note here that the ranking stipulated by k is irrelevant: it is only the ranking of various forms of i relative to the *satisfaction* of k that is at issue.) This is consistent with the stipulation that the pastor wants you to repent sooner rather than later, as long as that claim is read with an appropriate *ceteris paribus* clause. For it is true that, everything else being kept fixed, earlier repentance is to be preferred to later. If so, then $i_{prophet}$ is semipreferable to i_{pastor}. Assuming that this structure holds generally, the prophet's imperative is more urgent than the pastor's, even though both are satisfied in exactly the same conditions. Hence, for similar or identically satisfied imperatives, one can still compare their priority by triangulating against *other* possible actions one might undertake.

8.2.3 Advantages of the Account

This approach to imperative priority has several advantages. The best one is, of course, that imperative priority ends up depending on content. Further, it depends on content in a principled way. As I noted in chapter 5, the inclusion of $\gtrsim$ is necessary to capture the differences between commands that are satisfied in the same worlds. Those are not uncommon, and the difference seems to be one of content. So we need $\gtrsim$ anyway. Second, the account does so using a purely ordinal measure, $\gtrsim$, which leaves open the possibility that the extension to pains might treat pain intensities as merely ordered, rather than as a measurable quantity. Of course, if pain intensity were undergirded by a real-valued quantity that's intercomparable with other real-valued quantities, then $\gtrsim$ falls out trivially. But we can construct $\gtrsim$ even without some underlying magnitude.

The theory on offer also explains why imperative priority makes the most sense in limited domains—or, conversely, why many imperatives have *incommensurate* force. If I tell you to give money to Oxfam and your department head orders you to repaint your office, then there doesn't seem to be a well-defined sense in which one of those commands is more urgent than the other. That falls out nicely from our theory. The two imperatives are mutually selfish: each ranks its own satisfaction worlds higher than worlds in which it fails to be satisfied and the others are. So neither is semipreferable to the other. It is also possible for two imperatives to be incommensurate because each is semipreferable to the other with respect to different third imperatives. Such cases may play a more interesting role in action deliberation when the third imperative is not merely hypothetical but one that the agent must also try to satisfy.

Many simple imperatives are mutually selfish. When imperatives are mutually selfish, they don't contain information in their content that determines which should take priority. Of course, you must still make that a decision about which to do first. When you do that judgment will depend on facts extrinsic to the imperative—your other plans and desires, the source of the commands, and so on. That means that priority can be sorted out even in cases where urgencies are incommensurate. The sorting arguably has little to do with content. That might explain why imperative priority has received so little treatment in the literature: because it has been assumed to be a wholly extrinsic matter. Even then, selfish imperatives carry *some* information relevant to ranking—they still rank their satisfaction worlds after all. That information isn't useful in many cases where we have to aggregate across different imperatives. Further, there *are* important cases where two imperatives are commensurate, and where the more urgent one should thus be (all things considered) satisfied before, and in preference to, the less urgent one.

This take on incommensurability also preserves another important intuition. Although aggregation might be a tricky problem in general, it has a simple solution when the same action will satisfy (or tend toward satisfying) both imperatives. The library tells me to return a book, and the department head tells me to repaint my office. I may not be sure which is more important, and the two imperatives might be incommensurate. But going to campus is a necessary condition for fulfilling *both*, and so I have a reason to go to campus even though I'm not sure which thing I have reason to do first.

Finally, the present theory can account for cases where imperative intensities appear to have a cardinal structure. Suppose $\gtrsim$ has relatively fine-grained rankings over worlds, such that (for example) it distinguishes worlds where you satisfy i and forgo one unit of some good from worlds where you satisfy i and forgo two units of some good. Suppose further that we can identify points of indifference where satisfying i is as good as some quantity of x, better than any lesser quantity, and worse than any greater quantity. Given this, suppose we have two imperatives i and j, each of which has this structure relative to the xs, and further that i ranks its satisfaction as on a par with n units of x while j ranks as on a par with $2n$ such units. Under such circumstances, we can say that j is not just more urgent but *twice as urgent as i*.[3] One would, of course, need such a structure to be in place consistently—but if it were, it would be natural to speak of ratios of urgencies, not simply ordinal comparisons between them. Note here that the relevant ratio measure has no relation whatsoever to properties, but only to the rankings of different satisfaction and non-satisfaction worlds by the imperative. Hence, there are complex differences in the priority of an imperative, even if there is no sense in which that imperative somehow tracks a real-valued magnitude in the world.

8.3 An Ordinal Account of Pains

The extension to an ordinal conception of pain intensity is straightforward. Pains are imperatives. As part of their content, they have an especially rich and complex $\gtrsim$, one rich enough to make the sorts of comparisons noted above. So, for example, one pain is more intense than another just in case their world rankings agree that one should tend to the body part involved in the first pain at the expense of the one in the second.[4]

3 More precisely, under conditions where both rankings satisfy the axioms of the von Neumann-Morgenstern theorem. If so, then the theorem would ensure that we could find a suitable cardinal function that described an agent's behavior, and that would in turn be enough to account for ratio judgments of intensity as well as standard bets over merely likely satisfactions of i. If you don't like the arbitrary good x, then you might also construct a cardinal function out of those standard bets. Again, this is consistent with, but does not imply, the presence of an actual real-valued magnitude that gives rise to $\gtrsim$ (Ellsberg, 1954).

4 Another reason, albeit a tenuous one, to like the *body* as a source.

That commands *can* vary in intensity is, we take it, obvious enough. Further, the extension to pains seems straightforward enough: pains are imperatives, like other imperatives they can vary in intensity, and that variation has exactly the same sorts of practical consequences. A more intense pain in your foot will, all things being equal, cause you to forgo more activities to tend to your foot, to tend to your foot more urgently, and so forth. That's easily captured by the present account. A more intense pain will rank its satisfaction as higher than other things you might do.

This structure explains differences in intensities of pain. It provides a common structure in which other imperative sensations might be judged, while allowing us to account for the apparent oddity of fine-grained cross-modal intensity judgments. Again, I can judge the intensity of my pain and my hunger. Further, I might judge that my hunger is more intense than my pain. It is not clear, however, whether we're ever in a position to judge that some hunger is *twice as intense* as a pain.

If you agree that such judgments are problematic, then that problematic state can be accounted for by our view. Only within modalities does $\gtrsim$ have the requisite structure to make ratio judgments. Between-modality imperatives have enough structure to $\gtrsim$ to allow ordinal but not ratio judgments. Hence, cross-modal comparisons could be merely ordinal, whereas intra-modal ones remain cardinal.

Further elaborations of the model could account for other pain phenomena. Here is one intriguing possibility. Merely ordinal rankings—even ones that can be used to derive cardinal ranking functions—are not easily aggregated. Even if, say, the pain in my feet is more urgent than both my hunger and my thirst, it's not obvious whether I should stop hiking or continue on given that I am *both* hungry and thirsty. Models of aggregation do exist, however. In those models, different ordinal rankings receive weightings that determine how much they matter to the final aggregate.

In the context of pain, those weights might be interpreted as marking individual differences in the importance of pain: individuals who tend to assign low weights to pains will, all things considered, discount even relatively intense pains. Hence, this also gives a model for individual differences in imperative priority, one that would further distinguish among different imperative modalities.

8.4 Conclusion

In summary, pain intensity can be captured as part of the content of pains. It can be captured by appealing to details of the ranking function that forms one half of any imperative's content, and that can be generally used to account for intuitive differences in the priority of imperatives. Imperativism, then, is compatible with the idea that the intensity of pain can be determined by the content of the pain alone.

With that, I conclude the account of pain's content. I've shown that, under a plausible account of imperative content, the content of pains is sufficient to account for both the general (protection) features of imperatives and the main dimensions along which pains vary. This does not exhaust what could be said about the content of pains, of course. For one, individual pains might have even more complexity to them. I hope I've shown that the approach to content is general and powerful, hopefully enough to capture any further details you might please. For another, I've not said anything about the *instantiation* of this content in neural form. To reemphasize, this is an abstract theory of imperative content, one that (ultimately) must be augmented by some mapping to the biological facts. That will be a complex task, one that can't yet be completed, and which belongs to science rather than philosophy. By showing the utility of the abstract account, however, imperativism lays the groundwork for such a naturalization.

9 Objections, Replies, and Elaborations

9.1 Introduction

Imperativism explains a lot. There are odds and ends that need to be cleared up. In the next several chapters, I'll defend imperativism against objections that have been leveled against it. This will also elaborate and explicate imperativism's commitments. The objections can be grouped into three rough classes. First, one might worry that imperativism is *incomplete*: that is, that it doesn't capture some pains, or facts about pain. Second, one might worry that imperativism is *shortsighted*: that is, that the arguments for imperativism overlook some other more plausible motivating state that pains might be analyzed in terms of. Finally, one might worry that imperativism has actual *counterexamples*: that is, that there are cases of pains that simply do not motivate, which should be impossible on the imperativist position.

I'll deal with objections in that order. This chapter will handle a variety of putative counterexamples, each of which appears to show that imperativism leaves something out. Chapter 10 will examine and dismiss other competing views, and chapters 11 and 12 will deal with the specter of counterexamples to motivationalism.

9.2 Objections from Odd Cases

Imperativism is meant to be a theory about all pains. Some fit the mold well. The theory was crafted with pains of recovery in mind, and it also handles pains of acute injury and pains of warning.

The world of pain is diverse, however, and there are numerous odd cases to account for. Some of these pains are problematic for any account of pain. Others present a *prima facie* problem for imperativism in particular. Nevertheless, they must all be accounted for.

I'll lump these cases into two different categories. First, there are pains that have a connection to protection but that appear to be maladaptive rather than helpful. Second, and more troubling, there are pains that appear to be associated with parts of the body over which you have no control (in some cases because they don't exist) and thus can't do anything about. The first class presents no real problem. The second requires more care but brings out a distinctive virtue of the imperative account.

9.2.1 Maladaptive Pains

Pain is overall a good thing, and it helps keep us alive. But that doesn't mean that every pain is adaptive. As I noted in chapter 3, the imperativist is not committed to the idea that *every* pain promotes bodily well-being, any more than every hunger or itch does. Rather, the idea is that the pain *system* is constructed so as to produce imperatives that promote well-being in most cases, mostly reliably, in most circumstances, for most people. It is liable to malfunction, as is any biological system.

Nevertheless, maladaptive pains might seem puzzling on the imperative account because what they command is not immediately obvious to philosophical reflection. However, I'll argue there is usually no problem with specifying a plausible imperative *content* for a maladaptive pain. Indeed, in such cases, that content explains how the pain is maladaptive: obeying its command does not promote bodily integrity.

The most obvious cases are pathological pain syndromes. For example, allodynia is a condition where pain is provoked by stimuli that don't ordinarily cause pain: light touch, pressure, and mild warmth or cold. Allodynia is straightforwardly maladaptive. Yet there's no problem in specifying an imperative content for the pains of allodynia. Suppose a patient feels intense, burning pain when lightly touched on the arm. That pain commands protecting the arm, just as *your* pain would if a hot poker were actually pressed against *your* arm. The allodynia patient doesn't gain anything by complying with this command. But the content of the pain, and what it commands, does not differ from ordinary pains. A similar story can be told for the overenthusiastic pains of hyperalgesia. There, what has gone wrong is the intensity with which the pain is felt rather than the appropriateness of the pain. But the intensity motivates the sufferer to act just the same as an equally intense pain felt by an ordinary sufferer.

The same strategy can be applied to more common maladaptive pains. Most headaches, I'll assume, are nonfunctional. Many headaches are due to either muscular tension in the upper torso or poorly understood vascular changes in the head (perhaps caused by dehydration) (Sacks, 1986). Why *these* changes cause pain is, again, a question for the physiologist; I'll assume that headaches are just a classic case of a misfire in the most adaptive system. The most adaptive system might have reason to cause pain due to transient ischemia (perhaps to regulate exertion), and it might be unusually sensitive to transient ischemia even though the head muscles aren't involved in vigorous exercise. One has only so many options for building a body when you start from a single cell each time. That process apparently spreads receptors far and wide, including places where they're less useful. So it goes. Even if headaches are functionally mysterious, however, their content isn't: "Protect your head!"

Referred pains are a different way in which the pain system might malfunction. A standard signal of a heart attack is a strong pain in the left arm. Pressure on the sciatic nerve where it leaves the spine will result in pain in the lower leg. The physiological causes of referred pains may be complicated. But as I noted in chapter 7, the phenomenology is not. The pain caused by sciatica is (more or less) identical to the pain caused by some disorders of the foot itself. Both pains command you to protect your foot. Again, what's commanded has become unrelated to the underlying physiology. But the content of pain is unproblematic: it remains a protection imperative.

A final, trickier class of maladaptive pains involve those where the satisfaction conditions of the imperative would make sense in some context but are confusing or out of place in the actual contexts where the pain occurs.

One possible example in this class is menstrual cramps.[1] Primary menstrual pain is caused by the release of endometrial prostaglandins, which increase uterine tone and the strength of uterine contractions (Coco, 1999; Deligeoroglou, 2000). Prostaglandins also play a role in childbirth by causing uterine contractions, and they are sometimes used in that role to induce labor. Further, there's a correlation between severity of dysmenorrhea and severity of labor pain, suggesting that the same mechanism plays a role in both (Melzack, 1984).

So here is a plausible story about menstrual pain. The imperative that constitutes them is of roughly the same sort as the imperatives of childbirth—something about protecting the uterus, say. That one might feel such pains during childbirth is (let's suppose) both understandable and adaptive. However, what one is commanded to do by menstrual pain is irrelevant and non-adaptive.[2]

1 Menstrual pain is a complex topic, and the story I'll give is partly speculative. My account applies only to primary dysmenorrhea, not to secondary syndromes caused by underlying pathology. I focus on local cramping pain of uterine origin. It's possible that there's an additional contribution from passage of material through the cervix, as nulliparous women tend to have more pain. However, dysmenorrhea is associated with ovulation, not menstruation per se (Deligeoroglou, 2000). Further, both pain sensitivity and pain thresholds vary substantially across the menstrual cycle (Riley et al., 1999). This effect is not localized, suggesting a broad sensitization process.

2 Indeed, it is in some sense precisely backward—as (I've heard) the best way to alleviate cramps is to do things like stretch or exercise, which go against what the pain inclines you to do.

Put this way, however, the structure is familiar. There is a pain that commands something, but that something is actually quite different from what the body actually needs in that case. So pain promotes something maladaptive, or at least useless, rather than some more adaptive activity.

In general, maladaptive pains pose no deep problem for imperativism. In each case, the content of the pain can usually be determined by considering the content of some other, adaptive pain. In adaptive cases, we can see why following the imperative would promote health—and that is good reason to think that there is imperative content involved in the first place. That content in a different context might be maladaptive. Following pain's command might be useless or even counterproductive. Nevertheless, there is little doubt that maladaptive pain motivates the same kinds of *action* as a similar ordinary pain. The allodynia sufferer not only feels light touch, as if it were a hot poker pressed against the skin, but he also pulls back from that touch in the same way.

9.2.2 Unsatisfiable Pains

A second class of problem cases is far more serious. My attention was drawn to them by Maura Tumulty.[3] Consider phantom limb pains. As noted in chapter 7, amputation occasionally results in the persistent hallucination that the missing limb is still present (Ramachandran and Hirstein, 1998). These phantom limbs can be quite painful.[4] Of a friend whose arm had been amputated due to gas gangrene, W.K. Livingston writes:

> I once asked him why the sense of tenseness in the hand was so frequently emphasized among his complaints. He asked me to clench my fingers over my thumb, flex my wrist, and raise the arm into a hammerlock position and hold it there. He kept me in this position as long as I could stand it. At the end of five minutes I was perspiring

3 See Klein (2010) and Tumulty (2009). I depart a bit from her treatment, however. Tumulty's objection was directed at my (2007), which focused on motion proscription. There, I suggested that pains of the deep viscera—like the pain associated with kidney stones—might be understood as general imperatives to avoid motion of the torso. That cannot be right on a proscriptive account. At best, this would make renal colic an imperative against doing something *to* the kidney. If pains are imperatives against motion, then pains should weigh against doing something *with* the kidney. The point about renal colic is less pressing on a protection-imperative account: "protect your kidney" is arguably a satisfiable command. However, Tumulty's general point—that imperativism permits unsatisfiable imperatives—still stands.

4 Note that painful phantoms can occur even in the congenitally limbless, and so in those who have never had experience of moving a hand (Ramachandran and Hirstein, 1998). The historical frequency of phantom limb pain is a matter of some controversy. See Crawford (2009) for an intriguing argument regarding this issue.

> freely, my hand and arm felt unbearably cramped, and I quit. But you can take your hand down, he said. (Quoted in Melzack, 1973, 53)

I have claimed that pains are imperatives to protect a body part. The pain of a phantom limb must therefore be an imperative to protect (say) your hand. But on reflection, that's awfully puzzling. How, exactly, is it possible to protect a hand you do not have? Nor is there any closely related voluntary activity that could be proscribed against (as there was with menstrual cramps and bladder cancer). Nor will appeal to indeterminate content help: the problem is that no more determinate action can be performed with the hand. As Tumulty (2009) puts it, "Here, the problem is not (only) that pain may persist after a negative imperative is obeyed. Rather, the problem is that it is not clear what the relevant imperative would be" (165).[5]

Such cases are odd, but imperativism has something to say about them. True, Livingstone's friend could not protect a hand he didn't have. That does not mean that "Protect your hand!" is *unintelligible*, however. It is a perfectly intelligible command. It simply *cannot be obeyed*. The pain of a phantom limb, like a normal cramp, proscribes against continued action—say, against continuing to clench the hand. The sufferer has no hand, so he is ordered to do something impossible. Yet his pain has the same content as ordinary, nonpathological hand cramps. Some pains can weigh against actions that we cannot perform, but that does not present a problem for imperativism.

I think the appearance of a problem might stem from a misleading analogy with deontic claims. For claims about what you *ought to* do plausibly imply claims about what you *can* do. "You ought not keep doing that with your hand" is either absurd or false if you don't have a hand. If I claimed that this was the content of pain, I would be in trouble.

Imperatives do not work like deontic claims, however. Imperatives can be meaningful and motivating without being actually satisfiable. As unlucky privates learn in boot camp, one can be ordered to do something impossible, and that order can carry the same legitimacy as a satisfiable order. We are often told to do things that turn out to be impossible when we attempt them. It seems to me that any adequate semantics will have to treat some unsatisfiable imperatives as intelligible and legitimate. The model of imperative content sketched in chapters 5 and 8 were designed to preserve that intuition, and I'll return to it in section 9.3.3.

5 Tumulty's example is uterine contractions; for the reasons I've given above, I suspect that they require a slightly different treatment.

Phantom limb pain is an especially dramatic example of chronic and pathological pains. What is common to many such debilitating pains, I suggest, is that they present commands that cannot be satisfied. This would go some way toward explaining the frustrating, demoralizing effects of chronic pain.

I've not covered all possible strange cases of pain. But the preceding answers jointly sketch a general strategy for the imperativist to follow when confronted with such cases. First, does the pain motivate something that seems roughly useful? Then treat it like headaches or gout: the content of the pain is precisely what it motivates, and there's no further issue. If that fails, then can we tell a story (as we can in the case of menstrual cramps or cancer) about the underlying cause and what *would* be motivated in closely related circumstances? If so, that's a plausible guide to the imperative content. (Note that this works especially well for pathological pain syndromes such as allodynia.) Finally, for some pains, we must drop down to quirks in the implementation of the pain system (as with phantom limb pains) and look there for imperative content. In the latter cases, the imperative content may be literally unsatisfiable. Nevertheless, sufferers are motivated to *try* to satisfy the imperative, and their activity can be seen as evidence of such attempts.

In general, cryptic, strange, or otherwise unsatisfiable pains present no problem for the imperativist. So long as a sufferer is motivated to do *something* by their pain, that something is a good clue to a plausible imperative content. The only really problematic case, then, would be pains that don't seem to motivate at all or that motivate in odd ways; I will return to those in chapter 11. Otherwise, when a pain motivates, there's always a plausible imperative content lurking nearby.

9.3 Pain and Error

9.3.1 Error and Biological Role

I've argued that pain is conducive to survival. That does not mean that *every* pain is conducive to survival. Many clearly aren't. Many injuries give rise to chronic, maladaptive pains. Pains can be more intense than they need to be (say, paper cuts). That might seem like an objection to my view: how can I claim that pain is conducive to survival when many pains aren't?

This is a familiar situation. Many biological systems malfunction. The assertion that a system malfunctions is made relative to a notion of correct or proper function, and such ascriptions are in turn made relative to the overall survival value of the biological system in question (Neander, 1995). So too with pains. What helps survival is to have a pain system that more or less reliably commands us to protect ourselves in cases where protecting ourselves promotes bodily integrity. That system doesn't have to be perfectly reliable to be useful. Further, it is only because there *is* a determinate sense in which the pain system conduces to survival that we are able to talk about pathological pains as *mal*functions.

This is a point familiar in the pain literature since Descartes' discussion of phantom limbs in the *Meditations*. Descartes (2006) considers the possibility that God could have provided us with a pain system that gave us more complex or sensitive information, but that

> ... nothing else would have served so well the maintenance of the body. Similarly, when we need something to drink, a certain dryness arises in the throat that moves the nerves in the throat, and, by means of them, the inner parts of the brain. And this motion affects the mind with a sensation of thirst, because in this entire affair nothing is more useful for us to know than that we need something to drink in order to maintain our health; the same holds in the other cases. (49)

We are constructed with a pain system that, on balance, promotes survival. One of its survival-conducive properties is the relative *simplicity* of imperatives as an action-guiding tool.

That simplicity can go wrong, however, when the underlying system is compromised in various ways. I have argued that the best way to build such a system doesn't require conveying information about the negative states. A system of commands is just fine, and often preferable, to a system of indicative signals.

As I emphasized, that system can malfunction, and so motivate when it isn't appropriate, or do so more intensely than is required. So, for example, a paper cut might be extraordinarily intense and motivating, far more so than one would need to protect the cut. The most adaptive pain system is vulnerable to malfunctions, especially if it errs on the side of caution. That is a common enough situation. The eyeblink reflex is still useful even though it fires off dozens of times in error for every time it actually protects your eye (Millikan, 2004). Similarly, pain can be generally useful even though it often goes wrong.

Of course, as pain limits movement and so imposes a survival cost, we should expect unnecessary pains to be more rare than unnecessary eyeblinks—that's why misfires such as paper cuts strike us as *odd* cases. The mere fact that pain can malfunction, however, is no argument against the proposition that pains are homeostatic sensations.

9.3.2 Error and Disanalogy

A second way to run this objection tries to posit a disanalogy between pain and other homeostatic sensations. The best expression of this line comes again from Maura Tumulty. Tumulty argues that there is a disanalogy between pains and other imperative sensations such as itch, hunger, and thirst. Suppose (with Hall) one thinks that an itch says, "Scratch here!" (Hall, 2008). Scratch the itch, and it dutifully disappears. Not so with pain. The pain of a sprained ankle has a content such as, "Do not put weight on that ankle!" Yet my coddled ankle still throbs: obeying the imperative does not extinguish it. Tumulty (2009) thus notes correctly that, "Many pains persist, and persist at the same level of intensity, even when one obeys their commands" (162). If this is right, something has gone wrong: either I've miscategorized pains or else pains have some additional content that explains their persistence. For otherwise we must say that pains have gone wrong somehow, and that seems like a mistake.

The fact that pains persist even on compliance with their commands isn't a problem, however. As I argued in chapter 5, pains are best understood as conditional standing imperatives. Standing imperatives remain in force even as they are complied with. So if there is a problem, then it lies not in the nature of imperatives but in the apparent disanalogy between pain and other homeostatic sensations. Tumulty (2009) claims that, "In typical cases, it suffices, for a feeling of hunger or thirst to cease, that one comply with the imperative—that one eat or drink an adequate amount" (162). This is not quite correct. As I emphasized in chapter 2, what removes the imperative is not satisfaction per se, but rather the elimination of the underlying physiological cause of the imperative. The two coincide in ordinary cases: the satisfaction is a way of eliminating the underlying cause.

Pain is therefore precisely analogous to the other homeostatic sensations. In all cases, a homeostatic sensation promotes activity that eliminates the underlying physiological cause of the sensation under typical conditions. Thirst is often eliminated more quickly than pain. Rehydration is quick, whereas bones heal slowly. The disanalogy between pain and other sensations correctly reflects the differences in the underlying physiological processes on which these sensations depend. It need not reflect a fundamental difference in the type of content.

9.3.3 Error and Misrepresentation

A third objection stems from the fact that some pains seem to *mis*represent. A classic example is again phantom limb pains.[6] On one way of looking at them, phantom limb pains are *hallucinations*. Pain in a hand that does not exist seems like a perceptual mistake, in the same sense that the tactile hallucinations accompanying phantom limbs are perceptual mistakes (Tye, 1995; Hardcastle, 1999). But talk about pain hallucinations presupposes that pain has at least some truth-apt content. If it is possible for pains to be wrong, then it seems like pains must have some truth-apt content. Representational content fits the bill well. By contrast, phenomenal commands such as, "Don't put weight on that ankle!" are neither true nor false. Imperatives just aren't like that.

Note that this is a related but distinct point from claims about the maladaptiveness of certain pains. If you have the intuition that pains can be nonveridical in some sense, then maladaptive pains are surely among the nonveridical ones. Maladaptive pains, however, were supposed to pose a difficulty for the biological story that grounds imperativism.

Here, the problem is with the *content* of pains. If there is a sense in which pains can be nonveridical, then it seems that they must have at least some nonimperative content. Dual-aspect accounts can handle this easily, of course: the representational content of pain is the part that can be true or false, and the imperative content is the part that can't. Insofar as there is a problem, it arises only for pure imperativism.

One solution for pure imperativism would be to focus on implication rather than content. For example, even if imperatives are not truth-apt, we may still *infer* true propositions from *the fact that* a given imperative has been issued. Such inferences are possible (and reasonable) even if the content of pain is entirely imperative.

So imperative accounts might claim that patients are inclined to infer false propositions on the basis of phantom pains. This is implausible, however. Phantom limb sufferers do not have the auxiliary beliefs required to support such an inference—they are usually quite certain that they *are* missing a limb. (That's what makes phantoms so frustrating.)

I think, however, that there's a more sophisticated sense in which imperatives might link up to truth-apt content. That is to focus on the *presuppositions* of imperatives. Take the phantom pain described by Livingstone's friend. Its

6 Similar themes are explored in my (2012).

content is the imperative: "Don't keep clenching your fist!" Although neither true nor false, this imperative *presupposes* several false propositions: that there is a fist to unclench, for starters, and that it is possible to unclench that fist. These presuppositions are both false for the amputee.

More formally, consider the set of satisfaction worlds W_i, with which we have identified as part of the content of an imperative i. A *presupposition of* i is any proposition that is true in each $w \in W_i$. If I order you to get me a soda from the fridge, then I presuppose that there is soda in the fridge: every world in which you get me a soda is one in which there are sodas to be gotten. Similarly, if I tell you to keep the fridge from going empty, then I suppose that you can enforce restraint, *or* find more soda if necessary, *or* whatever: every world in which we don't run dry is a world in which at least one of these disjuncts comes to pass.

Every imperative presupposes some specific things about the world. Phantom pains presuppose something false about the sufferer's body. A cramping pain tells me not to clench my hand. Every world in which that imperative is satisfied is one in which I do something with my hand aside from clenching it. So the imperative presupposes that I *have* a hand. That is false in the case of phantom hands. So the pain of phantom limbs presupposes something false and is in that sense misleading.

All nontrivial imperatives also share a structurally similar presupposition, to wit, *that they can be satisfied*. Because W_i is defined as the set of worlds in which i is actually satisfied, the proposition that i can be satisfied must be true in each W_i world. This means that every imperative *presupposes* that it can be satisfied. Call the implication that i can be satisfied the *basic presupposition* of an imperative.

Phantom pains go wrong in a second way. As I noted in section 9.2.2, many imperatives are actually unsatisfiable—for example, because of simple lack of ability on the part of the addressee. You order me to lift that bus above my head. I can't. Nor can any of my close counterparts, no matter how hard they try. So the imperative is actually unsatisfiable. This means that the basic presupposition—that one can satisfy the imperative—is also false. So too with phantom pains.

In conclusion, there are two ways in which the commands that constitute phantom limb pains and the like can be said to imply falsehoods. First, in some cases, they presuppose that some body part exists, and that presupposition is false. Second, in all cases, such commands presuppose that they can be satisfied, and that is false.

Nevertheless, although they *presuppose* something false, they aren't actually false because imperatives aren't truth-apt. Thus, the pure imperativist can accommodate the intuition that something about phantom limb pains is false while also preserving the intuition that pains themselves are not truth-apt.

9.4 Pain and Introspection

Separating out the different aspects of pain experience, and showing how each depends on the content of pains, fleshes out the defense of pure imperativism. It also resolves a persistent puzzle for naturalists about introspection on pains.

What do we find when we introspect on our pains? The question is of some importance for intentionalist theories. The intentionalist, recall, claims that the phenomenal properties of experience supervene on their content; there are no additional noncontent-based intrinsic properties of experience of which we are aware. A standard argument for intentionalism is through the purported "transparency" of experience.

Transparency arguments claim that when we introspect, we only ever find facts about the intentional objects of perception, not facts about perceptual states. In a classic statement of the view, Gil Harman (1990) writes,

> When Eloise sees a tree before her, the colors she experiences are all experienced as features of the tree and its surroundings. None of them are experienced as intrinsic features of her experience. Nor does she experience any features of anything as intrinsic features of her experiences. And that is true of you too. There is nothing special about Eloise's visual experience. When you see a tree, you do not experience any features as intrinsic features of your experience. Look at a tree and try to turn your attention to intrinsic features of your visual experience. I predict you will find that the only features there to turn your attention to will be features of the presented tree. (39)

The relationship between transparency and intentionalism is contentious (for a good discussion, see Stoljar, 2004). These general worries noted, there is traditionally thought to be a specific, serious problem for intentionalism about *pains*. If the intentionalist line is right, this objection goes, then experiences of pain must also be transparent. Indeed, most intentionalists have assumed that pains pick out something in the world—some species of damage or disturbance—that we are transparently aware of when we turn our inner eye to pains (Harman, 1990; Byrne, 2001).

Many have objected to this move because it seems to be at odds with what we find when we actually introspect on pain. Murat Aydede (2001, 2006a), for example, has argued that when we introspect on pain, we are primarily aware of the pain, not any purported intentional object.[7] When we look inward at our pains, this line goes, we just find the pain, not anything about the body. Others have argued that the presence of intentional objects allows for an appearance-reality gap that doesn't seem possible in the case of pains. We do not normally speak of pain as the sort of thing that is veridical or nonveridical (Block, 2006). If a pain is there, then, we might say rhetorically, it is *there*: it strains common sense to speak of true and false pains.

This intuition may be partially due to a commitment to the incorrigibility of pain reports. For example, Baier (1962) claims that,

> When we talk about our pains, we are not cautiously talking about our wounds, or about our pain behaviour, for it is a mark of all these that they can be publicly confirmed or corrected, whereas it is a mark of pain talk that it cannot. On the contrary, the sufferer can say, with almost incontestable finality, whether and where he has a pain. (2)

But note that incorrigibility is really a distinct issue: incorrigibility accepts that pain reports *are* true or false but denies that we can sincerely make a false one. The current objection, by contrast, is that pains do not have the sort of content that can be spoken of as true or false. Similarly, it's not the same as claims of privileged access to mental states—privileged access applies to all mental states, whereas this objection is meant to apply specifically to pains and similar sensations. In a well-known passage, Kripke (1981) similarly argues for the lack of an appearance-reality gap in the case of pain, claiming, "To be in the same epistemic situation that would obtain if one had a pain *is* to have a pain; to be in the same epistemic situation that would obtain in the absence of a pain is not to have a pain" (152).[8]

Note, however, that these objections—and especially the ones that lean on an appearance-reality gap—presuppose that pain's content must be indicative. Such arguments lose their grip if we think of pains as having imperative content. What am I aware of when I introspect on a pain? Surely at least three things: where the pain is, what I should be doing with that body part, and how urgently I should be doing it. Those are all features of the *content* of pains. They are not experienced as features of the world. Well and good: they *aren't* features of the world! Commands reveal ways I must make the world be, not

7 Byrne (2001) discusses and defends against several variants of this sort of move.

8 Kripke's point is meant to apply to all sensations; I think it's telling, however, that he chooses pain—for it's there that people have the strongest intuitions about the lack of an appearance-reality gap.

ways it already is. When I look inward, those features of the command are precisely what I am aware of.

Of course, when I introspect, I might also be aware of a different fact: that my pain *hurts*. As I argued in chapter 4, this is a fact *about* pain, not an intrinsic feature *of* pain. So when I introspect on a pain, the things I find are partitioned into two classes: features determined by the pain's content, which is compatible with intentionalism, and features of other, distinct mental states, which have no immediate bearing on intentionalism at all. Once we spell out the specific content of pains, we find that pure imperativism is perfectly compatible with the transparency of perceptual states.

10 Why Not Some Other State?

10.1 Introduction

I have suggested that pains are naturally understood as states with imperative content. A key feature of imperatives is that they motivate. Many other mental states motivate as well, though. To conclude this portion of the defense, I suggest advantages of imperativism versus assimilating pains to other motivational states. I'll consider several possibilities in turn, and show the disadvantages of each. Having done so, I'll conclude by turning an eye to so-called *dual-aspect* theories of pain. Most theories of pain are actually dual-aspect theories, so it will be useful to examine the advantages of a single-aspect theory like pure imperativism.

10.2 Why Not Judgments?

Perhaps pains are something like a *judgment* that certain actions ought to be done, or should be done, or would be good to do. This is a rare view, although Nelkin's (1986) attitudinal treatment of pains arguably comes close. On the imperative account, hunger says, "Eat!" On a judgment account, it says, "You should eat," "Eating is good," or something along those lines. Rather than a command, then, perhaps pain is some sort of normative judgment.

Four disanalogies come to mind. First, it's contentious whether those sorts of normative evaluations actually motivate, especially in the direct and reliable way that homeostatic sensations need to motivate. Motivational internalism is a contentious view, whereas it's obvious that pains motivate. Second, these sorts of judgments require (in ordinary cases) a complex conceptual apparatus. Such an apparatus is not obviously available to children and lower animals that plausibly have homeostatic sensations. Third, judgments typically incorporate a fair bit of what you know and want; unlike lower-level mental states, they are not wholly encapsulated. Yet we can hunger and thirst while knowing full well that to drink would be bad for us. Fourth and finally, it's not obvious that judgment has any distinctive phenomenology. Insofar as it might, ordinary judgments seem unlike the sensory phenomenology of pains; they have a cognitive rather than a perceptual phenomenology.

Judgment views are thus unpromising. In section 10.6, I'll consider the most plausible alternative to judgment theories, which takes pains to be a perceptual analog of judgments rather than judgments themselves.

10.3 Why Not Desires?

Judgments seem too high-powered. What about desires? They definitely motivate, and they seem like the sorts of things lower animals have. So why not treat thirst as just a *desire* for a drink?

I think this ultimately falls prey to similar problems as the judgment account. Richard Hall (2008), in his defense of imperativism about itches, puts the problem well:

> Clearly hunger isn't simply identical with the desire to eat (or the desire for what is dry and hot), since one can have these desires without being hungry (think of a recovering hospital patient who is not hungry at all but whose doctor has told him in no uncertain terms that he must eat if he is to get well—and he wants to get well). Similarly for thirst. At most, hunger and thirst are particular *kinds* of desire to eat and drink.... The trouble with this kind of view, whether it be for hunger and thirst or for itches, is that it requires us to posit desires too far out on the periphery. It makes the senses too cognitive. Desires, along with beliefs and other propositional attitudes, are central. Your senses don't have them. Your senses just convey representations (and commands!) to you. You then believe (and obey) them or not. For example, you usually believe what your eyes tell you. But your sense of sight doesn't itself have beliefs. And your bodily sensory systems don't have desires. (532)

I think this is exactly right. Desires and beliefs are personal-level mental states. They are something that *we* have and that partly constitute us. That differentiates them from sensations, which are felt as things that happen *to* us. We don't have much control over our sensations, they needn't link up with our other desires, and they come to us unbidden. Hunger, thirst, and the other homeostatic sensations also come in that same unbidden way and are similarly beyond our control (that is why they are frequently annoying and inconvenient). Sensations are part of the more basic, peripheral *milieu* to which more central propositional attitudes must respond. We're better off looking for another sort of state.

10.4 Why Not Emotions?

One might treat homeostatic sensations as species of *emotion*. Emotions sometimes feel like they come unbidden and can also be at odds with our more central desires. For this line to work, we must choose the analogous emotions carefully. Hunger seems very much unlike complex, cognitively laden, world-responsive judgments such as grief and joy. If there is an analogy, then it is with putatively simpler emotions such as fear. Basic forms of fear seem more *triggered* than evaluated, and so they seem available to simple animals. If you're fond of neo-Jamesian theories of emotions, then emotions such as fear might also have a felt bodily component. Indeed, both A.D. Craig (2003a) and Derek Denton (2006) have written about "homeostatic emotions" and "primordial emotions," respectively.

I sympathize with Craig and Denton's accounts. I actually think we're on about the same thing, and that all three of us are appropriately sensitive to the motivational and biological role of the homeostatic sensations. Insofar as we disagree, the disagreement threatens to become merely verbal.

Nevertheless, I think that imperatives form a more useful category than primitive emotions. For starters, I doubt that "emotion" picks out a natural kind (Griffiths, 1997). I realize that is contentious; what is, I take it, less contentious is that "emotion" as a label has been applied to an extraordinarily heterogenous collection of mental states. There may be something in common between those states (I am suspicious of that myself). But even if there is, the utility of appealing to emotions as a philosophical *explanans* is relatively limited. For any useful story will have to explain what the particular emotions *are*. That is both difficult and philosophically contentious precisely because the word "emotion" has been used in so many different ways, including everything from merely triggered states to complex, evaluative, proposition-directed attitudes such as grief.

In contrast, imperativism sits on a simple, firm foundation. Like other sensations, pain is content-bearing. Unlike other sensations, that content is of a sort that is intrinsically motivational.

Indeed, I think the current objection can be turned around. Rather than worry about what sensations such as hunger have in common with emotions such as grief and joy, we are better off treating the "primitive" or "primordial" emotions as imperatives. Imperatives don't drag in extra baggage. Many theories of the emotions posit that they are essentially *valenced*: that is, that they feel good or bad. As I argued in chapter 4, the homeostatic sensations needn't be valenced in this way. I've suggested in passing that all of the homeostatic sensations may be profitably treated as imperatives, and it is precisely the set of homeostatic sensations that Denton and Craig want to account for.

10.5 Why Not Affordances?

David Bain (2011), responding to imperativism, suggests that even if pains specify behavior, they may not be imperatives. He suggests that,

> It might alternatively be, for example, that pain experiences simply *inform* us of what behaviour is such that, if it is not performed, injury will ensue. (A model might be the idea that visual experiences represent such "affordances" as the "jumpability" of a gap.) So the idea of pains as commands needs motivation. (179)

Given a recent resurgence of interest in Gibsonian notions of perception, this does seem like a possibility that might need to be ruled out.[1] Theories of affordance perception typically focus on perceiving facts about the external world rather than the internal *milieu*, but I'll assume there's some useful recasting of the idea to accommodate pain phenomena. The details won't matter much, for I think affordances aren't a good fit for pains. The problem with treating pains as affordances, I suggest, is that they do not bear the right relationship to motivation. Affordances permit, but do not require, certain kinds of actions. Pains do something more. That's reason to think that pain is not a perception of an affordance.

Following Chemero's (2003) formulation, affordances are relations between an ability of an organism and the environment. Affordances are best thought of as properties such as edibility. To say that an apple is *edible* for me is just to say that there's a relationship between my digestive abilities and properties of the apple such that the former can be exercised on the latter. To perceive an affordance is precisely to perceive this relational property. I'm happy to grant for the sake of argument that we can do so, and further that we can perceive relationships without being aware of the constituent *relata* (Chemero, 2003).

To perceive an affordance is thus to perceive an *opportunity* for action. Affordances do not directly *cause* actions. As Edward Reed (1996) puts it,

> The theory of affordances, on which all of ecological psychology is based, holds that the environment *affords* action for the organism, not that it *causes* action or that it *stimulates*

1 For varieties of the general Gibsonian revival, see Chemero (2009), Nanay (2011), and Siegel (2014). See also Matthen (2005) for an action-oriented view of certain perceptual qualities. Aside from Bain's suggestion, I'm not aware of anyone who actually defends the idea that pains are affordances. Siegel mentions imperativism in passing, but her concern is not with pain as such. There is work on the link between pain and negative affordances (e.g., of dangerous objects; see Anelli et al., 2013), but this is usually exteroceptive rather than interoceptive perception.

action.... Affordances provide opportunities for behavior and awareness. Whether the animal takes up these opportunities or not is a separate matter. (108)

I see that many things in my environment are edible. I don't eat all of them. In fact, right now I'm not eating *any* of the edible things I perceive. Why? Because I'm not hungry. So the presence of an affordance doesn't necessarily motivate: how I act toward an affordance depends on further facts about what I want.[2]

Pains aren't like that. What's important about pain is not that it presents an action as avoid*able*, but that it actually motivates avoidance.

Affordances do bear an interesting relationship to imperatives, and I think that close relationship may mislead. If affordances are opportunities for action, then the perception of an affordance might be the perception of something like a *permissive*. The difference between the perception of an affordance and that of a homeostatic sensation, then, is roughly the difference between "You may have a piece of cake" and "Eat this cake!" The former permits but does not oblige; the latter obliges and thereby motivates. Perceived affordances and imperative sensations may even work together: the combination of "Eat something!" and "This apple may be eaten" might be enough to motivate eating. But the content and role of pains and of perceptions of affordances are different enough that the two cannot be assimilated.

10.6 Why Not Felt Evaluations?

10.6.1 The Strategy

The final strategy I will consider—and the one that seems the most promising alternative—is what I'll call the "felt evaluations" strategy, after Bennett Helm (2002; see also Bain, 2013b). The felt evaluation strategy takes something like judgments and treats them as *perceptual* rather than cognitive states.

The following sentences seem roughly equivalent:

1 Don't put weight on your ankle!

2 Chemero (2003) suggests this point in passing, noting that affordances can't be relations to dispositional properties of organisms because dispositions are guaranteed to manifest when appropriate conditions are present. Chemero is concerned with the possibility of abilities failing even in appropriate circumstances, but I think the point can be generalized to show that affordances may or may not manifest depending on further facts about the organism. Whether this is compatible with the eschewal of internal states by ecological psychologists is an exercise left for the reader.

E Putting weight on your ankle is *bad*.

O You *ought not* put weight on your ankle.

Each will have same motivational effect (at least to a first approximation). Yet the first is imperative, and the latter two are indicative. The felt evaluations strategy, as I understand it, models pains to have indicative and normative content, similar to either **E** or **O**. Pains are sensory states, and sensory states with indicative content. What distinguishes pains from sensations in representational modalities such as vision is therefore not the *category* of content (all have indicative content) but rather the more specific contents themselves (descriptive vs. normative). The italicized portions in the example above emphasize the normatively laden content that distinguishes pains.

Some care must be taken when interpreting these putative indicative rephrasings. One can easily read sentences such as **E** or **O** as a form of thinly veiled imperativism. Language is complicated, and commands can be expressed by syntactically indicative sentences. To say, "Perhaps you should close the door" is, in the right circumstances, to say the very same thing as, "Close the door!" Call this route, by way of dismissing it, *British imperativism*. Expressing a command more politely does not make it less of a command. So if the felt evaluation strategy works, then it must be because pain sensations have some distinctively normative content, not merely that imperative statements can be rephrased in a vaguely indicative-looking way.

Viable positions would take seriously terms such as "the right thing," "should," "is good," or other similar possibilities. That is, the rephrasings would be syntactically indicative because they're meant to indicate something about the world: the *goodness* or *fittingness* or *rightness* of certain actions.

So, for example, Helm (2002) claims that pains are *felt evaluations*, that is,

> commitments that both are passive responses to attend to and be motivated by import and are simultaneously constitutive of that import by virtue of the broader rational patterns of which they are a part and which they serve to define. (19)

Similarly, Bain (2013b) focuses on the evaluation of bodily disturbances as being bad for you in some way:

> A subject's being in unpleasant pain consists in his (i) undergoing an experience (the pain) that represents a disturbance of a certain sort, and (ii) that same experience additionally representing the disturbance as bad for him in the bodily sense. (S82)

Pains then end up as something like a phenomenally salient sort of evaluation. Unlike simple judgments, they are perceptual (or quasi-perceptual) states. Unlike emotional theories, they give a *sui generis* account of what pains are (indeed, Helm's account grows out of a distinctive theory of the emotions, rather than taking emotional states to be basic).

The felt evaluation strategy requires that states with distinctively normative content have both a phenomenology and a motivating force. Both are potentially contentious; the latter in particular relies on a controversial motivational internalism. I'll put both of these claims aside, however. For even if we think that the idea of a felt evaluation makes sense, I think there are good reasons to prefer imperatives as the contents of pain. The objections go slightly differently depending on whether the indicative rephrasing is in terms of the *good* or in terms of *ought*-statements—that is, whether pains are more like **E** or like **O** above.

10.6.2 Pain and the Good

Treatments such as Bain's (2013b) evaluativism treat pains as representations of something *as bad*. The recommended content attribution to pain experiences ("this [actual/probable] tissue damage is bad for you") is not far from an outright recommendation to deal with and fix the tissue damage in question. The motivation, however, is presumably in terms of the normative content intrinsic to pains. Bain (2013b) argues that the reliance on normative content helps the evaluativist "make sense… of unpleasant pains' status as reasons" (S82). The idea, I take it, is that pains should *rationalize* behavior, not merely cause it, and that appeal to the good can have this rationalizing force.

I have already spelled out the sense in which pains have reason-giving force. But I suspect that there is a serious disanalogy between my account and the evaluativist. Imperativism treats the reasons that pains give as analogous to the reasons that practical authorities give. One might (perhaps even must) believe that accepting a certain source as a practical authority is a good thing. But that doesn't imply that to judge that *particular* commanded action is good to do. (Or that it isn't, for that matter. The goodness of particular actions is entirely besides the point.)

The evaluativist, by contrast, thinks that a reason for ϕ-ing must be a reason for taking it to be good to ϕ under the circumstances. The strategy has some prima facie plausibility: the claim is that if pain tells you not to put weight on your ankle, then it must be telling you that putting weight on your ankle is *bad*. It's a short hop from badness to action.

Put this way, the strategy is a more general instance of what has been called the "Guise of the Good" strategy in moral psychology. The guise of the good thesis says, roughly, that no one can be motivated to ϕ without apprehending ϕ-ing as good (in some sense). Apprehending of ϕ-ing as good is supposed to be what separates motivation from mere compulsion; equivalently, it's what gives proper motivational states their reason-conferring force. This is one reason, I suspect, why Bain and others are attracted to evaluativism: because they think there's a basic link between pains and the badness of particular states.

I think that evaluativism falls prey to the same objections that are ordinarily levied against the guise of the good thesis. Mark Schroeder (2008) notes that the guise of the good thesis comes in a strong and a weak variety. The strong variety says that motivational states must represent some action *as* good: that is, goodness must be part of the content. That's what evaluativism requires and what I deny. For representing something *as* good—and perceptually good—seems like a tall order. It's difficult to see how babies and animals can have the requisite concepts to have pains. The required concept of goodness, note, must be a fairly heavy one—for one gets to the guise of the good only by supposing that a fairly heavy notion of the good is required to confer reason-giving status on a motivational state. But that seems altogether too strong a requirement for such a basic state as pain.

The weak reading of the Guise of the Good thesis claims only that motivational states "aim at" the good. This would make Goodness the formal object of states such as pains but would not require pains to actually represent goodness as such. In this weaker sense, motivational states bear the same relationship to the good as indicative states bear to the truth (Schroeder, 2008). Saying "Grass is green" is not to say anything about truth as such, although of course it's part of the aim of assertions to say true things. Similarly, pains might aim at the good without having anything about goodness as their content.

This much weaker sense of the thesis, however, is entirely compatible with imperativism. It says only that imperatives aim at the good, in the sense that a system of imperatives is, more or less, good-promoting if followed. On that notion of good-promoting, I claim, providing adequate (good, rational) grounds for action can be and is customarily done both with imperatives and indicatives. States can be motivating and reason-giving without explicitly representing their objects as good.

10.6.3 Pain and Ought-Claims

A related view—although not one, to my knowledge, extant in the literature—would claim that pains have contents which contain something like an *ought* statement. This would retain some of the motivational attraction of evaluativism while distancing it somewhat from the problems inherent in the Guise of the Good thesis. The pain in your ankle says, "You ought to keep weight off of that." (Note that this "ought" must be read as a modal or else we are back to British imperativism.) Of course, even on my account, it is typically true that you ought to keep weight off of your ankle. That is made true by facts about you and your body. The view at issue is whether such a judgment *constitutes* the sensation.

I think that this recast version is no better than standard evaluativism. First, it falls prey to the same sorts of worries: normative modals seem too complex for the nonhuman organisms that can feel pain. Second, I think it sounds odd to say that homeostatic sensations present their corresponding actions as something that you ought to do. Sometimes that's just false: I can crave a cigarette, for example, while knowing and *feeling* that smoking would be entirely unpleasant and bad. Often goodness seems irrelevant. When I'm hungry, I don't typically feel like eating would be intrinsically good *or* bad—I feel like it's something I'm motivated to *do*, but the judgment comes later, if it comes at all.

Third and finally, it seems plausible that *ought* implies *can*. If a sensation delivers the claim that one ought to ϕ, then it should be possible for you to actually ϕ. If pain says that you ought to protect your hand (or do anything else with respect to it), then that requires you to have a hand. If you don't, then the resulting ought-claim is barely intelligible and certainly shouldn't be motivating. Yet phantom limb pain seems to motivate just as strongly as ordinary pains. This is puzzling if pains are normative judgments. If pains are imperatives, however, phantom limbs are perfectly intelligible, and the pain they present is straightforward. It may be *inconvenient* to be ordered to protect a limb you don't actually have, but that simply reduces to a command to do something that you can't in fact do. We get those kind of commands frequently. As I argued in chapter 9, unsatisfiable imperatives can still be motivating.

10.6.4 Pain versus Hurt, *Redux*

I have spent some time on evaluativism because I think there *is* something appealing about it. I just don't think it's attractive as a theory of *pain*. Instead, it seems to me that evaluativists have mostly been trying to account for what I've called *hurt* or *painfulness*. This is clear enough from their own quotes, where they are clear that they care about "pain" in the sense of felt *badness* of the particular state. Consider:

"Emotions are pleasant or painful—they feel good or bad—precisely because, as felt evaluations, they are feelings of positive or negative import." (Helm, 2002, 19)

"[The] question is made particularly interesting by two crucial features of unpleasant pains: their badness and their motivational force." (Bain, 2013b, S69)

"I am however denying that the painfulness of pain consists entirely in the character of those sensations." (Korsgaard, 1996, 147)

Note that in each case, there's a focus on the *painfulness*, *badness*, or *unpleasantness* of sensations. Helm (2002) clearly views this painfulness as a qualifier of other states (such as emotions); Korsgaard (1996) seems to have a similar feature in mind. Bain's (2013a) careful hedge leaves open the possibility that there are non-painful pains, but his actual focus is clearly on explaining why "as typically, pains *are* unpleasant, in virtue of what is this the case?" (Bain, 2013b, S69).

Evaluativism, I suggest, is most plausibly aimed at something *other* than pain: it's aimed at the badness, hurt, or painfulness that many pains (and other sensations besides) have as a contingent feature. I will return to this idea in chapter 14 when I consider stories about painfulness. For now, however, I will assume that the explanatory target is really distinct, and so properly understood evaluativism is not a viable alternative to imperativism about *pain*.

10.7 Dual-Aspect Theories of Pain

So far I have considered alternatives to imperativism that reduce pain to a single kind of state. Pure imperativism is also what I'll call a single-aspect theory. That is, it claims that pain has only one essential part: the imperative content that constitutes it.

Most contemporary scientific theories of pain, by contrast, are *dual-aspect* theories. Typically, the two aspects are conceptualized as a sensory part and a motivational part. Consider, for example, the well-known definition of pain offered by the International Association for the Study of Pain (IASP). Pain, according to the IASP Task Force on Taxonomy (1994), is "An unpleasant sensory and emotional experience associated with actual or potential tissue damage, or described in terms of such damage." The conjunction "sensory *and* emotional" suggests that when we have a pain, two things are going on at once.

Dual-aspect theories divide into two kinds. *Conjunctive* theories claim that pains are conjunctions of two ordinary mental states, each of which can and does exist separately but when conjoined constitute pain. Pain is thus, in Armstrong's (1962) lovely phrase, a "portmanteau-concept" combining two distinct mental states. A crude example of a conjunctive theory would be one on which pains are tactile sensations that we strongly dislike. Either tactile sensations or strong dislike might occur on their own; when they occur together, we get pain. A more sophisticated version of a conjunctive theory is Armstrong's (1962), on which pain is "*both* the having of a certain sort of bodily sense-impression, *and* the taking up of a certain attitude to the impression" (107).[3]

Contrast this with *composite* accounts of pain. Composite accounts claim that pains have two distinct aspects or properties, neither of which is typically observed on its own. The two aspects might be independently manipulable, but a state with just one of the aspects is either impossible or pathological. The typical composite view posits a representational aspect and an affective/motivational aspect, each of which explains distinct features of pain. Composite views are currently popular among both scientists and philosophers; even

3 I also read Kurt Baier (1962) as defending a conjunctive theory; he also gives conjunctive readings of several other theories at the time. The theory outlined in Tye (2006) is a more ambiguous case.

other imperativists about pain often include imperatives as only one of the two aspects of pain.[4]

Pure imperativism claims that pain has only one aspect—the command that forms its content. That's what it is to be a pain, and that's all it takes to be a pain, full stop. This must be something more than the claim that pain has only one essential aspect but is frequently *accompanied* by a wholly distinct mental state. For the latter would not be a dual-aspect theory: it would be one on which pain has a single aspect and a bunch of reliable but contingent followers-on. That is precisely the sort of view I've argued for.

Dual-aspect views are popular, however. If a dual-aspect theory is right, then pure imperativism is wrong. Are there good arguments for a dual-aspect view that remain untouched by the above?

Arguments for dual-aspect views fall into three categories. First, there are purely philosophical arguments. So, for example, one might argue that pains are sensations, sensations must have content, and therefore pains must have representational content. But because pains *also* motivate, in a way that representations alone don't, they must have some additional component. I take it I've shown the flaw in these sorts of arguments: imperativism demonstrates how sensations can be both content-bearing and motivating, and to do so in a way that preserves the biological role of pain.

Further, as I demonstrated in previous chapters, imperativism is compatible with pains having a felt location, a felt intensity, and a felt quality. Dual-aspect theorists typically locate these in the "sensory" part. I've shown that this isn't necessary. Further, by situating these qualities in the motivating imperative, they are properly connected to motivation. As I argued, it is not a mere *coincidence* that a pain in your finger motivates you to protect your finger, whereas a pain in your toe motivates you toward your toe. By linking the qualities of pain tightly to what pain *does*, imperativism gives a unified picture of pain and its role.

Insofar as there are philosophical arguments for a dual-aspect view, then, they must rely on more general positions about pain—that it must be representational, or that it must represent something as bad, or in general one of the many ideas that I have already considered. In most cases, I've suggested, these

4 On the science side, see for example Price (2000) and Price and Aydede (2006); for philosophers, see Dennett (1985) and Hardcastle (1997). Nikolai Grahek (2007), about whom more shortly, is the most strident contemporary composite theorist. He combines philosophical and psychological argument for a composite view. Richard Hall (2008) is a pure imperativist about itches but a composite theorist about pains.

alternative positions are intrinsically problematic; yoking two together won't overcome those problems.

The second sort of argument comes from empirical data in ordinary subjects (Fernandez and Turk, 1992; Gracely et al., 1978; Price, 2000). Studies of pain reports consistently show two dimensions of variation, commonly referred to as the "sensory" and "affective" dimensions of pain. These can be independently modulated via hypnosis (Rainville et al., 1999) and correlate with activation in distinct brain structures (Rainville et al., 1997).

These data, however, are entirely consistent with imperativism if we assume a distinction between pain and *hurt*. One need only posit—as seems plausible—that pain induced in a laboratory situation typically hurts, and that the "affective" dimension therefore tracks hurt rather than pain. Further, nothing in this research commits the "sensory" dimension to being representational: the typical mark of the sensory dimension is often simply that it is localized, which is consistent with an imperativist account. Recall from chapter 1 Hare's distinction between different ways that "pain" is used. There is a common usage where "pain" picks out *both* the sensation of pain and the accompanying hurt. The imperativist need only claim that this research used "pain" in the common sense, thus gathering up data on both pain and the accompanying hurt. That does not mean that hurt is a part of pain, however, or that it necessarily accompanies it.

Dissociating the two would also explain the otherwise puzzling results on empathetic and social pain.[5] What does having your finger smashed have in common with seeing someone else's finger smashed, or hearing someone laugh at you for having a smashed finger? They all *hurt*. Separating out that hurt from the more basic sort of physical pain should, I suspect, provide better perspective on both states.

The third sort of argument, and the most compelling if it worked, is from putative dissocations between the two aspects of pain (Grahek, 2007; Hardcastle, 1997). On the one hand, there are apparent cases of pain that do not motivate—cases such as so-called morphine pain and pain asymbolia. On the other, there are cases that are meant to be pain with hurt but without the localizable sensory component. If successful, this double dissociation would be a

5 For empathetic pain, see Singer et al. (2004); for social pain, see Eisenberger and Lieberman (2004) and MacDonald and Leary (2005). The suggestion is that each of these involves an other-directed state that hurts but is not a physical pain.

good argument for a dual-aspect view and a serious problem for pure imperativism. (Indeed, depending on how one parses the remaining components, it might be trouble for imperativism in any of its forms.)

With that in mind, the next chapter will turn to the evidence. Spoiler: I don't think it's actually very good. So I'll conclude that the evidence for a true dual-aspect theory is surprisingly weak. Pains can do all of what they need to do if they're just one thing—an imperative. They can do all of what we observe them to do if that imperative is often painful. To complete the defense, then, I'll consider putative cases of pains that do not motivate.

11 Pain Asymbolia and Lost Capacities

11.1 A Potential Counterexample: Pain Asymbolia

I fractured my ankle when I was thirteen. In the hospital, I was given a shot of morphine. Then, for some short and blissful period of time, something odd happened. I could still feel pain in my ankle. But I didn't really *care*. The pain that had motivated me so strongly moments before became irrelevant; it no longer bothered me.

So-called "morphine pain" is extremely common in emergency room situations.[1] Phenomena like it are troubling for imperativists like me. Imperativism is a species of what I'll call *motivationalism* about pain. Motivationalists claim that motivational force is an intrinsic property of pains. Many have denied motivationalism. The debate can be shaped by empirical facts. Find someone who is entirely unmoved by pain, and motivationalism is threatened.

Because imperativism is a motivationalist theory, potential counterexamples are especially pressing. My thirteen-year-old self felt pain, and imperativism says that he was feeling a command to protect his ankle—a command that he blithely and blissfully ignored. Commands should motivate, however. If they don't, imperativism is in trouble.

Morphine pain appears to be such a counterexample, and numerous scientists and philosophers have taken it as such.[2] Morphine pain is a bit tricky, however. We usually take reports from people on powerful narcotics with a grain of salt, and there are practical and ethical barriers to thorough experimentation in the emergency room. Narcotics also surely diminish secondary motivation: that shot of morphine also removed my fear and anxiety and so profoundly diminished my suffering.

The motivationalist can and should concede that emotions such as fear, frustration, and anger are only contingently connected to pain. What pains necessarily motivate are actions that protect our bodily integrity; other negative affective states depend on a cognitive evaluation of the significance of pain. A stronger case is needed to threaten motivationalism.

1 Thanks to Dr. Edward Thompson for valuable, if admittedly anecdotal, evidence in this regard. There are reports of similar phenomena caused by synthetic mu agonist opiods (such as Meperidine) that differ structurally from morphine, various barbiturates (Keats and Beecher, 1950), and nitrous oxide (Hall, 1989). This suggests that the phenomenon does not depend on some effect specific to morphine.

2 For a sampling, see Aydede and Güzeldere (2002), Beecher (1956), Dennett (1985), Hall (1989), and Tye (1995). Also tempted by this line, although they waver, are Pitcher (1970) and Stimmel (1983).

11.2 Pain Asymbolia

In a recent book, Nikola Grahek (2007) presents a much stronger apparent counterexample to motivationalism. This is the strange case of *pain asymbolia.* Pain asymbolia is a rare condition caused by lesions to the posterior insula (Berthier et al., 1988). Asymbolics say that they feel pain, but they are strikingly indifferent to it. In the first reported case, Schilder and Stengel (1928) note that,

> The patient displays a striking behavior in the presence of pain. She reacts either not at all or insufficiently to being pricked, struck with hard objects, and pinched. She never pulls her arm back energetically or with strength. She never turns the torso away or withdraws with the body as a whole. She never attempts to avoid the investigator. (147)[3]

Strange enough. But not only do pain asymbolics fail to react to such stimuli, they also appear to recognize what they feel *as pains*. Schilder and Stengel (1928) continue:

> Pricked on the right palm, the patient smiles joyfully, winces a little, and then says, "Oh, pain, that hurts." She laughs, and reaches the hand further toward the investigator and turns it to expose all sides.... The patient's expression is one of complacency. The same reaction is displayed when she is pricked in the face and stomach. (147)

As Schilder and Stengel (1928) note, their patient was in no way inattentive or unaware. Quite to the contrary, she was actively engaged with the investigators, and readily offered up new body parts to be poked and prodded.

Pain asymbolia is arguably the cleanest apparent counterexample to motivationalism. Asymbolics thus appear to feel pain without being motivated by it. There seems to be no overriding motivation that explains asymbolics' lack of response. Unlike lobotomized patients, asymbolics don't protect their bodies when they encounter stimuli that cause them pain. Unlike drugged patients in emergency rooms, there is a well-established tradition of using first-person reports of lesioned patients, and neurology permits detailed tests of their responses.

It is worth emphasizing just how strangely asymbolics behave, even outside of laboratory settings. Schilder and Stengel's (1931) patient would readily stab

3 Quotations from Schilder and Stengel (1928) are my translations from the original German. Aleks Zarnitsyn and Mae Liou helped with the translation.

herself with needles and jam objects into her eyelid. Berthier et al. (1988) report an asymbolic patient who suffered a severe burn at home because he made no attempt to escape the danger. Hemphill and Stengel (1940) note of a patient:

> The absence of any defence or withdrawal reaction was clearly shown when a strong, painful sensation was applied by surprise, e.g. when the examiner, standing behind the patient, suddenly pricked his hand or neck. When the patient was threatened with the first he made no effort to guard himself or to withdraw his head, nor did he show any instinctive combative reaction. (256)

More generally, pain asymbolics seem willing to submit to ghastly batteries of tests, even though many of these tests are actually injurious.

So much for motivationalism? I say, no. In this chapter, I'll argue that Grahek has misinterpreted pain asymbolia. Grahek treats asymbolia as a deficit of sensation. I'll present an alternative view, on which asymbolics have lost a fundamental capacity to care about their bodies. The alternative view better explains the wide variety of phenomena associated with asymbolia. Having argued thus, in the next chapter, I'll show that this capacity-based view is compatible with a moderate form of motivationalism, and I'll show that imperativism is committed to precisely that moderate form.

The argument in this chapter is pitched at a fairly general level; it's meant to show that motivationalism is plausible, within constraints. Of course, I hold a more specific version of motivationalism. As a preview, here's what I think the imperativist should say. In chapter 6, I emphasized that we take the body as a minimal practical authority. Doing so is a precondition for pain to motivate, although once we *do* accept the body as an authority, pain motivates directly. We accept the body as a minimal practical authority because we care about our bodies. For obvious reasons, that care is stable and constant, at least in ordinary cases. The asymbolic is not an ordinary case. He has ceased to care about his body and so has ceased to treat his body as an authority. That is what explains his strange condition. That is compatible with motivationalism, or at least the variety endorsed by imperativism.

11.3 Two Models of Asymbolia

Whether asymbolia represents a counterexample to motivationalism depends crucially on how we understand what's going on with the asymbolic. To begin, then, I present two models of asymbolia—Grahek's and my own—and defend the latter.

11.3.1 The Degraded Input Model

Here is one model of what has gone wrong in asymbolia. Pain is actually a composite mental state. It has (at least) two proper parts: a sensory part, perhaps representing tissue damage, and an affective/motivational part, which moves us to act. These two parts typically go together, and there is good biological reason for them to do so. Under the right conditions, however, one or the other can be absent.

This is a version of a dual-aspect theory. As I noted in chapter 10, variants of this composite view of pain are popular among both philosophers and scientists. Grahek (2007) also endorses it:

> although pain appears to be a simple, homogenous experience, it is actually a complex experience comprising sensory-discriminative, emotional-cognitive and behavioral components. These components are normally linked together, but they can become disconnected and therefore, much to our astonishment, they can exist separately. (2)

When the components of pain come apart, strange syndromes result. Asymbolia is a paradigm case. The pain of asymbolics, Grahek argues, has lost the affective/motivational component. As such, "[Pain] becomes a blunt, inert sensation, with no power to galvanize the mind and body for fight or flight. Such pain no longer serves its primary biological function" (Grahek, 2007, 73).

Call this the *degraded input* (DI) model of asymbolia. DI claims that asymbolics have a deficient sensation: their pain lacks the motivational push that our ordinary pains possess. This explains why asymbolics are indifferent to pain: the pain has changed. The DI model is incompatible with motivationalism. According to DI, the motivational force of pains comes from their affective/motivational component. That component can go missing, but the sensation remains a pain.[4] So motivationalism is false.

4 Note that not all composite theories are incompatible with motivationalism. One could argue that pain must have both components to be deserving of the name, or that "pain" properly refers to the motivational portion, not the sensory one. (Armstrong (1962) argues, for example, that pains are a combination of a tactile sensation plus an extreme dislike of that sensation.) Grahek (2007, 95), drawing on Hardcastle, eschews this strategy. I'm happy to follow his lead for the sake of argument. Grahek is not entirely consistent on this point: see his discussion of components of pain versus *real* pain at 111.

Grahek argues that there is a double dissociation between the sensory and affective aspects of pain. Pain asymbolia provides one half of the dissociation: as he puts it, asymbolics feel pain without painfulness (where "painfulness" refers to pain affect).

The other half of the dissociation—painfulness without pain—depends on the case described by Ploner, Freund, and Schnitzler (1999) of a patient with a unilateral lesion to SI and SII. Laser stimulation to the left (contralateral) hand did not elicit pain sensation but did produce in the patient a "clearly unpleasant" feeling that he "wanted to avoid." Grahek takes this as a case of pain affect preserved in the absence of pain sensation. Double dissociation[5] between two mental processes is usually taken as evidence that they are only contingently related (even if they typically occur together). So the composite view of pain falls out directly, and DI appears to be well motivated.

While the other half of purported dissociation—painfulness without pain—is less relevant to the question of motivationalism, I'm also suspicious of it. First, the patient described by Ploner, Freund, and Schnitzler (1999) *did* appear to feel pain in the hand contralateral to his lesion, albeit with a much higher threshold (see their figure 2). Further, their patient described the sensation he was feeling as "unpleasant" before he felt pain. But there are many unpleasant sensations aside from painful ones. Why think that the patient felt the negative affect associated with *pain* rather than just some other unpleasant sensation?[6]

11.3.2 The Lost Capacity Model

DI is not the only way to interpret asymbolia. Here is another model: Asymbolics don't react to pain because they no longer care about the physical integrity of their bodies. More precisely, they have lost the *capacity* to care about their bodies in whatever way is relevant to pain. They do not care about cuts, burns, and scrapes because they can no longer conceive of why such events are bad. Call this the *lost capacity model* (LC) of asymbolia. Both LC and DI predict that asymbolics will be unmoved by pain. They differ, however, on the

5 Grahek never uses the term "double dissociation," although his argument is obviously meant to be read as appealing to a double dissociation between pain affect and sensation. (For a more explicit double dissociation argument for the same conclusion, see Hardcastle, 1997.) The point is not merely pedantic. *Paired* dissociations are crucial bits of evidence; single dissociations are hard to interpret and provide weaker evidence for separability (Shallice, 1988). I will argue shortly that Grahek misinterprets the putative dissociation provided by asymbolia.

6 Similar remarks apply to Hardcastle's (1997) interpretation of tooth pulp stimulation under the influence of fentanyl.

explanation of that fact. DI says that something has changed about the sensation of pain. LC says that something has changed about the *person* not the pain. Further, LC predicts that the deficits in asymbolics should be relatively widespread. Asymbolics should be indifferent not just to pain but to *any* immediate threat to their bodily integrity. Information about such threats can come from a variety of sources: sensation, language, beliefs, and so on. Caring about the integrity of your body requires hooking up sensation, cognition, affect, and behavior in the right ways, regardless of how one comes to know about a threat. According to LC, asymbolics lack this integrative capacity because their lesion has destroyed the neural substrate on which the capacity depends.

The LC model faces an initial empirical complication that is worth addressing. If "threat to bodily integrity" is understood so broadly as to include the threats that come from failure to eat or urinate, then LC looks empirically false. Schilder and Stengel's (1928) patient, for example, asked to eat and use the bathroom. If, as LC claims, asymbolics aren't motivated to protect their bodies, then what motivates them in these cases?

Distinguish between *immediate* and *distant* threats to one's body. Avoiding an immediate threat requires action now or in the near future; avoiding a distant threat can be done on one's own time. Pains are associated with immediate threats. Hunger represents a distant threat: failure to eat will eventually cause damage, but one typically has considerable time and flexibility in choosing how to meet that threat. The need to urinate even more so: one has to exert considerable force of will to be damaged rather than merely embarrassed. Further, eating and bathroom-going are likely to be tightly regulated in the institutional settings in which most asymbolics reside. In general, counterfactual analyses work reliably only for immediate threats. Were *I* to stop eating now, I would likely end up in the hospital, not dead. Asymbolics' behavioral oddities seem to be limited to direct threats.

The lack of response to direct threats admits of several possible explanations. Here's one I find the most plausible.[7] Our responses to distant threats are largely shaped by habit. Most of us eat at fixed times, for example, and just

7 It is not the only plausible story. Caring about one's body in the case of immediate threats might just dissociate from caring about one's body in the case of distant threats. More generally, the insula is a complex and functionally differentiated structure that underlies many different interoceptive functions (Ibañez et al., 2010). So it is possible that damage might spare some functions and not others. That is an empirical matter, one complicated by the fact that insular damage is typically quite widespread and messy. Such dissociations would be surprising, but no more so than those found in other areas of neuropsychology. The conclusion would then be that asymbolics have lost the capacity to care for their bodies in some ways but not in others. This possibility is

because it is time to eat. Asymbolics' atypical responses to distant threats, then, may be accounted for by the retention of habits that promote bodily integrity even in the absence of the underlying capacity to care about bodily integrity. Responses to immediate threats are, for obvious reasons, much harder to shape and so much less dependent on habit. Hence, a lack of care would show up most obviously in responses to immediate threats—such as those associated with pains and the like. While both DI and LC are psychological-level theories, a brief note about the brain is in order. Both Grahek and I accept that asymbolia results from damage to the posterior insula, a cortical region plausibly involved in integrating sensory and limbic signals related to pain (Craig, 2003b). We differ on how to interpret the functional consequences of this damage. Drawing on a proposal first put forth by Geschwind (1965), Grahek argues that pain represents a *sensory-limbic disconnection syndrome*. On his view, damage to the insula in asymbolics prevents limbic processing from being incorporated with sensory processing (Grahek, 2007). DI is motivated by this picture: there are two processing streams in normal folks, one of which has become a dead-end in asymbolics.

Geschwind's model of disconnection syndromes has been criticized for assuming an entirely serial, feed-forward picture of the brain (Catani et al., 2005). On his view, each brain region performs a specialized function and passes on the result to higher association centers, which in turn pass on their results to still further association centers, and so on. Earlier processes in the causal chain are entirely unaffected by later ones. DI embodies a picture like this: the sensory deficits of asymbolics are caused by a failure of limbic processing to be attached appropriately to sensory processing in some later stage.

This simplistic model of brain function has fallen out of favor. The insula projects back to the limbic system and receives input from a variety of frontal areas. Thus, it seems to do more than simply composite together the results of earlier sensory processing stages—instead, it plays an active role in integrating multiple different cognitive processes, especially interoceptive and motivational ones (Singer et al., 2009). LC is partly inspired by this more complex picture of the insula.

A final difference is worth noting. DI treats the motivational force of pain (when present) as a brute fact about pain: some sensations just have the power

empirically distinguishable from the story I suggest, although I find nothing in the actual clinical literature that would allow us to do so.

to motivate, and pain is one. LC, in contrast, gives an explanation of just why pains motivate. Pains motivate because we care about our bodies. If we were to stop caring—something that's ordinarily impossible, for good biological reasons—then pains wouldn't matter. Asymbolics are a realization of this unusual possibility.

11.3.3 Evidence for a Lost Capacity

Both LC and DI predict the pain-related deficits of asymbolics. LC, however, predicts that there should be a general loss of appreciation for threats to bodily integrity. DI does not. The clinical literature supports LC. First, asymbolics are not indifferent to pain alone. They also appear to be indifferent to any dangerous or threatening stimulus. Hemphill and Stengel's (1940) patient was also "quite disinterested" when matches were struck close to his face and eyes, and he showed no response to unexpected loud noises or strong flashes of light. Schilder and Stengel report that their patient also failed to respond to (a) being threatened with a hammer, a knife, and a needle; (b) shrill whistles; and (c) a magnesium wire burned inches from her face (Schilder and Stengel, 1928).

Asymbolics' indifference is not limited to simple sensations. Berthier et al. (1988) report that five of their six patients failed to respond to "verbal menaces." Schilder and Stengel note that their patient "shows no appreciation at all for threats of pain *or for any threats in general*" (Schilder and Stengel, 1928, 154, my italics). Hemphill and Stengel's (1940) patient showed an unusually dangerous indifference to threat:

> The patient was observed proceeding one morning along the main road of the hospital. He made no effort to get out of the way of a lorry behind him in spite of the loud warning of the horn. That he heard the horn and recognized its character is certain, for he admitted as much with considerable heat when he was forbidden, for his own safety, to walk alone on the main road. (256)

LC handles these various phenomena well. It predicts that asymbolics should be indifferent to bodily threats *regardless* of modality.

What about DI? Grahek (2007) mentions these phenomena. He suggests that the relevant deficit is plurimodal and does not discuss the issue further. I can think of two readings of this suggestion, neither of which is terribly satisfying. First, Grahek could mean that asymbolics have a conjunction of many specific deficits. That is, asymbolics fail to attach motivational force to pain, *and* auditory sensations, *and* visual sensations, *and* to written and spoken language, and so on. Any of these deficits could in principle occur on their own; in asymbolics, they happen to occur together perhaps because of the anatomical proximity of distinct functional substrates. This interpretation is possible, but it seems *ad hoc*. It merely posits a distinct and potentially dissociable deficit for every modality that experimenters have thought to test, with no further positive evidence other than the deficits. Of course, the multiple-deficit version of DI might still be true; without further evidence, it is not convincing.

Second, Grahek could mean that there is a single deficit that manifests across a variety of sensory modalities. This would presumably be a conduction deficit: that is, the failure of a linkage between the limbic system and higher association areas. This is more plausible. However, it still requires a certain degree of special pleading. Asymbolics' deficits seem to be limited within modalities as well: they are indifferent only to sensations conveying bodily threat, not to sensations generally. Schilder and Stengel's (1928) patient, for example, had a strong emotional reaction to being called a liar and a thief. So her deficit cannot simply be one in attaching emotional valence to sensation and language quite generally: she is indifferent only to utterances that involve threats.

On either reading, DI faces a further difficulty. Suppose I anesthetized your arm and placed it out of sight. Suppose I then told you that I was pummeling it with a hammer. You would, I suspect, be motivated to act—to remove your arm, to flee, and to rethink your reasons for trusting me in the first place. *Why* would you be motivated? Not because of some sensation you're having. Your arm is insensate and occluded. Instead, you would be motivated by a simple bit of practical reason: you care about your body, caring about your body requires avoiding needless injury, needless injury is happening, and therefore you have a reason to act. So we can be motivated to protect our bodies in two ways: directly, because of some sensation we're having, or indirectly, because we believe that our body is being harmed.

What about asymbolics? By all accounts, they seem to lack *both* ways of being motivated. They are not motivated by their pain, but they also aren't motivated by *the fact that their body is being damaged*. That fact should be apparent to them if DI is true—both because they retain the sensory, informative aspect of pain, and also because they know what is happening to them. Again, asymbolics readily submit to actually injurious tests. Again, they are actually injured because of their condition. This is puzzling. If asymbolics lacked *only* the motivational aspect of pain, then we should expect them to be otherwise like us when it comes to bodily damage. But they are not.

A useful comparison is with the congenitally insensitive to pain.[8] From birth, the congenitally insensitive don't feel any pains at all. *A fortiori*, they don't have sensations with whatever affective/motivational component Grahek thinks is critical for pain behavior. Yet they still learn to protect their bodies as best they can. That is, they learn what situations are injurious, and they avoid these situations precisely because they don't want to be injured.

If we accept Grahek's account, then asymbolics' total lack of motivation is puzzling. Grahek claims that the pain of the asymbolic lacks the usual affective component, and that explains their lack of response. But if that was *all* that was missing, then we'd expect the asymbolic to be like the congenitally insensitive to pain: unmotivated by the sensation of pain but still motivated to protect their bodies when they learn of threats. On the contrary, the asymbolic appears to be entirely uninterested in the fate of their bodies, *however* they learn about an injurious situation. Hemphill and Stengel's patient who put himself in danger on the road did not react to the sound of the horn. But he also did not react to *the fact that a truck was bearing down on him.*

The point may be put in a slightly different way. The composite account of pain claims that asymbolics still have the sensory aspect of pain intact. What does that sensory aspect do? On most accounts, it informs about bodily damage or the like. (It could be a bare *quale*, but even then the presence of that *quale* is reliably associated with bodily damage and so provides useful information.) So according to DI, asymbolics should still know that they are being damaged. As per the bit of practical reason above, they should still be indirectly motivated to act. But they aren't. Grahek (2007), remember, says of the pain of the asymbolic that it is "a blunt, inert sensory appearance with no power to galvanize the mind and body" (73). But that would make the "sensory-discriminative" aspect of pain unlike any other sensations we're familiar with. The sensation of seeing blue doesn't have (in ordinary cases) a motivational-affective dimension. But it still does something: it informs us that there is a blue thing nearby. On Grahek's story, the sensory-discriminative function of pain appears to be wholly epiphenomenal. It is there. We can make verbal reports about it. That is the only causal consequence it seems to have for our behavior. That's deeply odd.

8 Grahek (2007), unlike many authors, correctly distinguishes asymbolia from congenital insensitivity.

In contrast, LC gives a perfectly straightforward story about asymbolics' general lack of concern. Asymbolics don't care about the integrity of their bodies because they can't. The capacity they lack applies to sensory evaluations of stimuli, cognitive evaluations of threat, and any way in which we might normally learn that our physical integrity is jeopardized.

Asymbolics often fail to make even reflexive responses to stimuli they describe as painful. Schilder and Stengel report that their patient made only mild reflexive responses to extremely intense stimuli and none at all to less intense manipulations. Several people have worried that a high-level explanation such as my own cannot account for this fact. Why would spinal-level reflexes be reliably suppressed by something like lack of care?

This objection depends on an inaccurate picture of spinal reflexes and their top-down control. As I noted in chapter 2, all spinal reflexes are continually modulated by top-down signals from the cortex. Further, many motor reflexes are altered in characteristic ways after frontal damage: stereotyped withdrawal reactions to injurious stimuli, for example, may become contextually inappropriate after severe brain injury (Plum and Posner, 2007; see Berthier et al., 1988, for a report of contextually inappropriate peripheral responses in asymbolia). Hence, new stimuli do not provoke otherwise absent downward modulation—they merely change an ongoing modulatory process. Top-down modulation doesn't need to wait for signals to arrive from the periphery in order to be effective. Further, many motor reflexes are altered in characteristic ways after frontal damage: stereotyped withdrawal reactions to injurious stimuli, for example, may become contextually inappropriate after severe brain injury (Plum and Posner, 2007).

Most work on downward modulation of nociceptive pathways focuses on sensitization. However, while relatively rare, some central abnormalities do seem to result in diminished or absent peripheral pain reflexes: the phenomenon is attested in cases of catatonia (Northoff, 2002) and schizophrenia (Dworkin, 1994). While speculative, I think the rough outlines of an explanation can be offered. Spinal neurons receive modulatory input from higher levels of cortex (Millan, 2002; J. Price, 2005). Asymbolics' lack of peripheral reflexes is likely due to a disorder in this complex system. The lack of care for one's body might directly cause abnormal tonic inhibition of reflexes. Reflex suppression might also be a mere side effect: perhaps some insular region inhibits a subcortical inhibitory region, and damage to that structure results in increased inhibition overall (compare in this regard the first stage of the Sprague effect; Sprague, 1966). In any case, the lack of peripheral reactions to pain is consistent with other evidence on downward suppression of spinal-level reflexes. In summary, then, lack of concern is entirely compatible with lack of ordinary "reflex" activity.

11.4 Conclusion

To conclude, there are two ways in which DI might be defended. The first is Grahek's, by arguing for a classic double dissociation between pain sensation and pain affect. Asymbolia is supposed to be one half of the dissociation—pain without painfulness. But asymbolia does not fit the classic double dissociation model. A dissociation requires severely impaired performance tasks involving one mental component and preserved functioning on other tasks. Asymbolics, however, do not behave as we would expect someone with a mere sensory deficit to behave. Their indifference runs deep. So there is no classic dissociation, and the argument fails.

Second, DI could be defended abductively—either as a neuropsychological argument or as a more general species of inference to the best explanation. The strength of an abduction depends on the power of competing hypotheses. I argue that LC is a stronger explanation of asymbolia. Some phenomena it explains directly, while DI needs complex or *ad hoc* hypotheses to account for them. Other phenomena are explained by LC but not by DI. So on balance, we have reason to prefer LC.

In summary, it appears that asymbolics have lost the ability to care about their bodies, and this lack of care explains their strange behavior. With that in mind, it's time to return to the question of motivationalism. Is the LC model compatible with imperativism?

12 Asymbolia, Motivation, and the Self

Morphine alters the whole cycle of expansion and contraction, release and tension. The sexual function is deactivated, peristalsis inhibited, the pupils cease to react in response to light and darkness. The organism neither contracts from pain nor expands to normal sources of pleasure. It adjusts to a morphine cycle. The addict is immune to boredom. He can look at his shoe for hours or simply stay in bed. He needs no sexual outlet, no social contacts, no work, no diversion, no exercise, nothing but morphine. Morphine may relieve pain by imparting to the organism some of the qualities of a plant. (Pain could have no function for plants that are, for the most part, stationary, incapable of protective reflexes.)
—William S. Burroughs (1957)

12.1 Three Kinds of Motivationalism

Suppose that the argument in the previous chapter is conclusive, and that the lost capacity model is correct. Is LC compatible with motivationalism? Unsurprisingly, that depends on how we understand motivationalism. More surprisingly, the answer is yes. There is a philosophically interesting version of motivationalism to which asymbolia is no counterexample.

First, assume that all viable forms of motivationalism are hedged in the ways considered in chapter 11. That is, when we say that an agent is motivated by pain, we mean they are disposed to perform certain actions to protect the integrity of the physical body (although that disposition can be overridden and need not result in further negative affective states). Given that, here are three ways to understand motivationalism about pain:

Ambitious Motivationalism: Necessarily, if an agent feels pain, they are motivated by it.

Modest Motivationalism: Pains motivate in virtue of some property *P*, and pains intrinsically and necessarily have *P*.

Lazy Motivationalism: If a typical agent in normal circumstances feels pain, they will be motivated by it.

Lazy motivationalism is not threatened by asymbolia. LC tells us that asymbolics aren't typical agents in typical circumstances. So there's at least one way of understanding motivationalism on which it is compatible with asymbolia.

That is an unsatisfying victory. I suspect that few have been tempted to deny lazy motivationalism and fewer still for good reason. DI, note, is entirely compatible with lazy motivationalism. Lazy motivationalism is thus too weak to capture a real debate. Most important, although lazy motivationalism might be descriptively accurate, it sheds little light on pain: in particular, it says nothing about *why* normal circumstances and typical agency matter. Let's put it aside and try for something stronger.

Ambitious motivationalism *is* philosophically interesting. Grahek denies it, and it does place some strong constraints on our theories of pain. It is arguably the most intuitive way of cashing out the motivationalist thesis. But Asymbolia is also clearly a counterexample, even if we accept LC. LC does not deny that asymbolics feel pain, nor that they are unmotivated by it. Ambitious motivationalism says that is impossible. So ambitious motivationalism is false.

That leaves only the carefully hedged, modest motivationalism. Like ambitious motivationalism, the modest variety is incompatible with Grahek's view, so it is philosophically interesting. Unlike lazy motivationalism, it makes a strong claim about the nature of pain: however pains motivate, they always have the property in virtue of which they do so. Modest motivationalism, however, doesn't claim that pains *always* motivate: just that they always have the property in virtue of which they motivate.

The key claim is that motivation is a two-place *relation* between a sensation and an agent: my pains motivate *me*. Modest motivationalism says that this relationship can fail to hold. If it does, however, it must fail in virtue of a change in the *agent*, not because of a change in the pain.

An analogy might help illuminate the matter. Both lit matches and chlorine trifluoride are ignition sources: they have the power to start fires. Chlorine triflouride will start fires (nearly) anywhere and on anything.[1] Lit matches, by contrast, start fires only if certain background conditions are in place: there must be oxygen and dry tinder, the air can't be too humid, and so on. Given these conditions and a lit match, a fire will start. We are happy to attribute to lit matches the property of being an ignition source despite this. This is because matches have the right sort of intrinsic property that causes fires to start; that distinguishes them from other things (bricks, donuts, puppies) that don't. If a struck match *doesn't* ignite, then we blame conditions not the match.

1 More precisely, chlorine trifluoride is hypergolic and an extremely strong oxidizer, and so will start fires in the absence of oxygen and in materials not normally thought of as flammable—sand, concrete, asbestos, water, and so on (Clark, 1972).

Ambitious motivationalism views pains as a bit like chlorine trifluoride: they light the fires of action come what may. Modest motivationalism, in contrast, says that pains are like matches. They always have an intrinsic power to motivate, but that power manifests only if circumstances are appropriate.

Of course, modest motivationalism runs the risk of collapsing back into lazy motivationalism. One needs to say more about the necessary background conditions. This is where the LC model of asymbolia does real work. The LC model suggests that the relevant background condition is the capacity to care about the fate of your body. That has gone missing in asymbolics and explains why they aren't motivated by their pains.

LC is a substantial empirical and philosophical claim. First, it claims that there *is* a unified capacity for caring about your body in the right way. The care we have for our bodily integrity isn't just caring about pains, and *also* caring about sudden noises, *and* also acting appropriately when you believe you're being injured, and so on. All of these more particular states are manifestations of a single general capacity and so must stand and fall together. That in turn has empirical consequences. Grahek claimed that pain asymbolia was a specific dissociation between pain and motivation. It is not. If LC plus modest motivationalism is true, then there cannot be any such specific, simple dissociation. Instead, any agent who is indifferent to felt pain should be as asymbolics actually are: possessed of a collection of deficits that manifest in many different but related ways. That in turn makes strong, falsifiable empirical predictions.

For the connoisseur of the neuropsychology literature, an aside. One might object that the above story is built on evidence from association of deficits. In his influential work, Shallice (1988) has argued that associated deficits are a weak foundation for neuropsychological inference. Two points are worth noting. First, Shallice's argument is strongest against syndromes posited on the basis of probabilistic generalization over groups of patients, which is not at issue here. Instead, the prediction is that distinct tests of the *same* construct will show similar patterns of impairment: that is, there is only one psychological deficit that manifests itself in various ways on various tests. Second, Shallice argues that associations of deficits are evidentially shaky, as they can always be overturned by dissociations observed in the future. That is true but also a *virtue* of the present account: it is empirically riskier and so easier to falsify.

12.2 Motivation and Command

Modest motivationalism is thus compatible with LC, and so with the empirical facts of pain asymbolia. Maintaining modest motivationalism requires, however, that pain have some intrinsic property *P* that motivates under ordinary circumstances. Further, *P* must persist, without motivating, in cases of asymbolia. The argument I gave was deliberately abstract and gives different types of motivationalism an opportunity to slot in their preferred property for *P*.

The plausibility of modest motivationalism ultimately depends on whether there is some appropriate property, and many otherwise plausible candidates don't fit the bill. *P* arguably cannot be the property of *representing* facts about damage (or other bodily states). The property of representing bodily damage doesn't seem to have the correct direction of fit to be motivating: if I come to learn that I have been damaged through some other route, then I may or may not be motivated to do anything about it. Nor does the property of being a desire, emotion, or other affective state seem to be a candidate for *P*. Asymbolics feel pain, but they do not appear to have any of the ordinary affective states associated with it. So even if typical pains *also* motivate in virtue of some associated affective states, asymbolics can recognize sensations as pains without those states. They are thus not candidates for an *intrinsic* property in virtue of which pain motivates. These objections are not decisive, and the clever motivationalist may find ways around them. A motivationalist could also treat *P* as some *sui generis* property with just the features required, although I suspect such a move will be philosophically unsatisfying.

There is one candidate for *P*, however, that I think is both philosophically satisfying and compatible with motivationalism. That candidate, of course, is an imperative. Imperativism clarifies how motivation might break down in cases such as asymbolia. As I noted in chapter 6, being motivated by a command requires accepting the source of the command as a minimal practical authority. While acceptance relies on some reason or other, once one *has* accepted a source as an authority, those reasons need not enter into further deliberation. But if the reason for accepting an authority breaks down, commands from that authority may cease to be motivating without a change in the content of the commands.

The reason why we accept bodily commands is because we care about our bodily integrity. In treating our body as an authority, we accept that it might sometimes make mistakes. Indeed, the nondeliberative nature of motivation by pain implies that we will continue to be motivated by bodily commands even when we *know* that a mistake has been made (as in the case of phantom limbs). However, if we were to cease caring about our bodily integrity entirely, then the authority of the body would be undermined. Pains would cease to motivate. That failure would occur without any change in the imperative content of pain.

This, I suggest, is what has happened in the case of pain asymbolia. The asymbolic recognizes pains because their ordinary imperative content has not changed. However, the asymbolic has ceased to treat bodily commands as authoritative and so has ceased to be motivated by them. The situation of the asymbolic is thus a bit like the unperceptive man who hears someone who is (in fact) a police officer shout, "Stop or I'll shoot!" He can recognize the utterance as a command. He might think that under the right circumstances, someone hearing such a command has a good reason to stop—all without realizing that those circumstances actually obtain.

The resulting picture is a form of modest motivationalism: it says that pains possess an intrinsic property (imperative content) in virtue of which they ordinarily motivate, while admitting that pains may in fact fail to motivate given suitable changes to the *agent*.

Pains motivate in virtue of their intrinsic motivational properties. That motivation occurs, however, only because we accept the commands of the body as authoritative. That opens the possibility that pains may, in special cases, fail to motivate protective action. Motivational failure requires a severe breakdown of the authority of the body—a breakdown that has other, wide-ranging consequences, and so does not constitute a counterexample to motivationalism properly construed. Hence, asymbolia and related syndromes do not constitute counterexamples to imperativism.

12.3 Asymbolia and Depersonalization

I suggested that asymbolics lack the capacity to care about bodily integrity. That doesn't need to manifest as an occurrent belief about lack of care. Instead, it may manifest itself more as a type of *indifference*. One's body becomes, as it were, just another object in the world. An odd object, perhaps, that still commands you to care for it—but not an object that you have any deeper reason to care for than anything else around you.

If this is right, then the phenomenology of asymbolia might resemble a kind of *depersonalization syndrome*. The Diagnostic and Statistical Manual of Mental Disorders, 4th edition (DSM-IV) defines depersonalization as "a feeling of detachment or estrangement from one's self" and notes that,

> The individual may feel like an automaton or as if he or she is living in a dream or a movie. There may be a sensation of being an outside observer of one's mental processes, one's body, or parts of one's body. Various types of sensory anesthesia, lack of affective response, and a sensation of lacking control of one's actions, including speech, are often present. (§300.6)

Perhaps asymbolics' experience of pain is an experience of a certain kind of detachment from that pain. They recognize it as pain, but in some important sense, it has ceased to be something worth caring about. It thus has the feel of a sensation that they can no longer identify with as their own.

(This might also explain why many asymbolics appear amused or befuddled by their pains. Ramachandran (1998), noting frequent reports where asymbolics laugh in the face of pain, argues that they recognize the incongruity between typical responses and their own. In support of this explanation, other asymbolics seem to feel the need to rationalize their responses. An asymbolic described by Hemphill and Stengel (1940) rationalized his absence of reaction to pain by saying, "I am used to that because I have worked on the road" and "Labourers are always hurting themselves; we don't take any notice of it" (256). All of these appear to be reasonable responses to an unusual depersonalization experience—for asymbolics can certainly remember that they *used* to be motivated by their pains, and that most reasonable people are, and be struck by this incongruity without being able to explain it.)

That damage to the insula might produce depersonalization is not surprising. There is a growing consensus that the insula plays a complex and active role in maintaining representations of the body, especially facts about homeostatic needs. One crucial function it plays seems to be in supporting what Craig (2002) calls *interoception*: that is, awareness and reflection on the state of one's body. Damage to the insula can produce a variety of deficits of bodily self-awareness (Ibañez et al., 2010). As Karnath and Baier (2010) note, this can include asomatognosia (the feeling that a patient's limbs do not belong to them) or somatoparaphrenia (the feeling that a patient's limbs belong to someone else). Damage to the insula thus seems to interfere with identification of sensations as our own.

Further, the feelings of disengagement typical of depersonalization can extend to the sensation of pain. Mauricio Sierra (2009) notes the similarities between asymbolics and the utterances of patients with depersonalization disorder.[2] One such patient remarked that while he felt pain, "it is as if I don't care, as if it was somebody else's pain" (Sierra, 2009, 49). Another patient, on being pricked with a pin, said that the sensation was "as if it were being done to another person" (150).

Depersonalization is also a symptom of other psychiatric diseases, including schizophrenia. Some schizophrenics are indifferent to pain, sometimes to the point of self-mutilation. Many authors assume that this phenomenon is due to schizophrenic insensitivity to pain.[3] However, recent reviews of the literature have noted that schizophrenics appear to have the same pain threshold as normal subjects, and that this effect is present even in unmedicated schizophrenics (Bonnot et al., 2009; Singh et al., 2006). Guieu et al. (1994) thus argue that, for schizophrenics, "the term of 'indifference' to pain may be more appropriate than 'insensibility' to pain" (255). Finally, and perhaps most intriguingly, Wylie and Tregellas (2010) have recently noted consistent evidence that abnormalities in the insula are often associated with schizophrenic depersonalization symptoms, and they suggest that the phenomenon may be similar to pain asymbolia.

Treating asymbolia as a species of depersonalization disorder is thus an intriguing possibility. For one, it means that asymbolia is not a *sui generis* deficit. It is instead a specific and severe form of a more common disorder, and one that those interested in pain might study more readily. That's handy: asymbolics are rare and difficult to study.[4] In turn, we might find analogues of depersonalized pain in even more prosaic situations, including those that have long intrigued philosophers writing on pain.

One such case, mentioned in the beginning of chapter 11, is that of morphine pain. Patients given an acute dose of morphine often say that they are indifferent to their pain. Morphine can produce powerful feelings of depersonalization. Conversely, patients with depersonalization disorder have compared it to the effect of morphine. Noyes and Kletti's (1977) patient, for example,

2 Relevant to the present discussion, Sierra (2009) also notes that purely sensory theories of depersonalization have long since fallen out of favor.

3 Including Grahek (2007).

4 Asymbolics often have severe language deficits. That is probably a neurological accident: the insula is located near important language centers, and the lesions that produce asymbolia are usually large.

remarked on a depersonalization episode, "I would compare it to a morphine 'high'; I once had morphine after an operation" (378). We might thus understand morphine pain as a variety of drug-induced depersonalization: patients are indifferent to pain not because the pain has changed but because they no longer appreciate it as a command worth following.

For what it's worth, my interest in the topic stems in part from my own experience with morphine pain. I find the above a phenomenologically accurate account of my experience. While the pain persisted for a bit, I could no longer see what it had to do with *me*—there was pain in my ankle, but what was going on down there didn't seem especially relevant. Of course, this is a self-serving, 20-year-old anecdote about a drug experience; the reader may want to take it with a grain of salt. Still, the powerful depersonalization that morphine produces, and the indifference that Burroughs describes to all forms of bodily care, fit my experience well.

Finally, treating asymbolia as a species of depersonalization might be relevant to current debates about the unity of consciousness. The asymbolic, and the depersonalized more generally, feel sensations that they are estranged from—that they do not take to be *theirs* in the sense that we normally do. This may not threaten some forms of the unity of consciousness thesis: there is another important sense in which the pain is their sensation whether they realize it or not. For example, it does not threaten something like what Bayne (2010b) calls phenomenal unity or what Rosenthal (2005) dubs the thin immunity principle.[5] However, it does show that there is another sense in which our sensations may be unified: as sensations over which we have a feeling of ownership. Asymbolia, and depersonalization more generally, show that this sort of unity may fail. Their failure comes not from a change in the sensations we feel but in the sort of agents we are. These syndromes show that failures of this kind of unity are not just real but have grave consequences.

12.4 A Contrast: Lack of Suffering

12.4.1 Setup

My defense of motivationalism focused on the primary motivational force of pain. That was the most pressing challenge: asymbolics didn't appear to be

5 Although for worries about the latter in the related phenomenon of somatoparaphrenia, see Liang and Lane (2009).

motivated *at all*, which was in tension with imperativism's claim that pains were constituted by commands. However, the pains of asymbolics clearly don't *hurt* them either. Depersonalization shows that it's not simply the sensation of pain *as such* that we find unpleasant. Rather, it is in part the motivational *effect* of pain that makes it unpleasant. When pains no longer move us, they hurt us less, if at all.

I'll talk about the connection between hurt and primary motivational force more in chapter 14. I want to walk a fine line, however. On the one hand, as I argued in chapter 4, pains without hurt are not only possible but actually occur in ordinary people. On the other hand, pains are typically painful, and pains of even moderate intensity appear almost inevitably so.

The distinction between the two states opens a possibility that might occur in unusual or pathological states. It should be at least *possible* to have even severe pains that motivate but from which we don't suffer. That is, it should be possible to have pains that retain their primary motivational force (and so strongly motivate bodily protection) but that don't have any secondary motivational force (and so the person who feels them has no particular inclination to get rid of them). I've already noticed that analogous cases occur more frequently for other homeostatic sensations. Often one is hungry and so motivated to eat but is not inclined to get rid of the sensation as such.

Strong pain without suffering is relatively rare. But I think it also occurs, and it is worth thinking a bit about the circumstances under which it does. To start, it's worth carving off a few sorts of cases that are only superficially similar to what I'm looking for but aren't actually cases of pain without suffering.

First, cases such as pain asymbolia aren't good test cases. If I'm right, then primary motivational force is also missing in asymbolia. So the real explanatory task for the imperativist isn't to show why the asymbolic's pains don't hurt but why they don't move them to protect their bodies (*contra* Bain's (2013a) critique).

Second, and similarly, we should be careful to look only at cases where there is pain. Some authors, impressed by Beecher's account of the soldiers at Anzio, have taken this to show that the unpleasantness of pain consists of a cognitive evaluation of the goodness or badness of the event (Bain, 2011; Hall, 1989). I've already noted that this is a questionable interpretation: ordinary folks can be injured painlessly in circumstances they judge to be entirely negative (Melzack et al., 1982). But in any case, it's clear from Beecher's account that his interviewees often didn't feel pain *at all* (Beecher, 1956). Hence the lack of primary motivational force is not odd.

Third, we should distinguish our target from cases where we've got pain along with accompanying *pleasantness*. I'll consider the most striking version of this, the masochistic pleasures, in chapter 13. For now, I'll note that masochistic pleasures typically involve states that also *hurt* and are described as such (Pitcher, 1970). Masochistic pleasures thus present their own puzzles but are not really what I am looking for.

12.4.2 Pain Without Hurt

Non-examples aside, I think a few plausible cases remain.

The effect of some drugs seems to be to reduce only the secondary motivational force of pain. This appears to be the case with lower doses of opiate analgesics and with the relative indifference to pain caused by large quantities of alcohol and other hypnotic drugs (Keats and Beecher, 1950). Of course, one runs the risk of confusing these states for the more powerful depersonalization states brought on by higher doses of similar drugs.

However, I think it's probably not a stretch to claim that the apparent indifference to pain seen outside of pubs is probably an elaboration of alcohol-induced indifference to life's woes. Note too that at least some of the drunk and injured do appear to be cognizant of their wounds and take steps to protect them, which differentiates them from the asymbolic. This is an area where the intrepid reader may be urged to do further fieldwork.

Lobotomized patients also appear indifferent to previously intractable pain. It is possible that their injuries reduce their care, and so make them like asymbolics. There is ample evidence that lobotomies produce a state of profound apathy, including apparent apathy toward one's bodily functions (Barber, 1959). However, as Melzack and Wall (1996) point out, these patients still withdraw from pinprick, avoid walking on broken ankles, and generally react to pain as we do. The pain of the lobotomized thus retains its biologically basic motivational force; what has gone missing are the other emotions usually associated with strong pains. Brand and Yancy note that, "Patients report feeling 'the little pain without the big pain'" (quoted in Grahek, 2007, 32), suggesting that what is missing are secondary reactions to pain, not its core motivational import. Even authors who emphasize the apathy of the lobotomized also note that they are able "to respond normally to nociceptive stimulation" (Barber, 1959, 439).

Pharmacological and surgical interventions are striking examples. Are there less drastic, nonpathological cases of pain without hurt? I think the clearest cases involve the effects of hypnotic analgesia. Hypnotic suggestion can effectively suppress many kinds of pain, in both clinical and experimental settings (Montgomery et al., 2000). It is more effective for acute pain than for chronic pain (Patterson, 2004), but for acute pain, it can be surprisingly effective. As Montgomery et al. (2000) note, the effect does not merely appear to be a reporting bias or desire for compliance on the part of patients. Patients will, for example, voluntarily receive less morphine after hypnosis. Hypnosis is also effective for pain control, even absent other analgesia, during horrifically painful procedures such as burn debridement and bone marrow aspiration (Patterson and Jensen, 2003). It is unlikely that those patients are faking it.

The effect of hypnosis on pain is a matter of some debate. At least some authors have argued, on neurological grounds, that the effect of hypnosis is similar to that of depersonalization syndromes (Röder et al., 2000). Typically, most follow Rainville et al. (1999) and argue that pain has both an affective and a sensory-intensive component, and that these can be independently manipulated by hypnosis (see Patterson and Jensen (2003) for a review). These results are usually interpreted within a dual-aspect framework. As I argued in chapter 10, the imperativist can easily reinterpret dual-aspect theories as being about the primary and secondary motivational forces of pain. That is, interventions that reduce "sensory" pain should be read as reducing the intensity of pain, whereas interventions that reduce "affective" pain should be read as reducing the secondary motivational force of pain—typically, the degree to which pain hurts. Because there is evidence that hypnotic intervention readily reduces the affective intensity of pain, we have some reason to believe that the hypnotic state is or can approximate pain without hurt.

Indeed, it is not terribly surprising that hypnosis and other cognitive interventions might at least *reduce* hurt. For hurt comes in degrees as well, and the intensity of hurt is only partially dependent on the intensity of the underlying sensation. The same twinge in my arm might barely register if I think it is a cramp, but it can cause deep distress if I think it is angina. Many post-exercise aches hurt worse if you think they are due to injuries rather than merely pains of recuperation. This suggests that the felt badness of pains might be a more thoroughly cognitive phenomenon—or at least that it might incorporate personal-level beliefs and desires more widely than pain does. This leads to the final place in which reduction of suffering might be achieved. At least some suffering comes from the way in which we fight pain and are nevertheless overcome by it. As Korsgaard (1996) notes, "Pain really is less horrible if you can curb your inclination to fight it. This is why it helps, in dealing with pain, to take a tranquilizer or to lie down" (147). If we could give up on that fight, at least a bit, then we might reduce our suffering.

One potential mechanism for this might be found in various forms of mindfulness meditation. Shinzen Young (2004), a meditation teacher who writes on pain management, claims:

> For most people, the notion of pain that is not suffering may sound like an oxymoron, a contradiction in terms. Most people have trouble imagining what the experience of pain without suffering would be like. Does it hurt? Yes. Is that a problem? No.
>
> People have difficulty understanding this because they are not familiar with the experience of pure pain, that is, pain without resistance. Since much of our habitual resistance to the flow of pain begins at a preconscious level of neural processing, by the time we consciously experience a wave of pain, it has already been frozen into suffering by unconscious resistance. In other words, most of us cannot remember experiencing pure pain. What people call "pain" is actually a combination of pain and resistance. (41)

Young claims that the psychological struggle against the command of pain constitutes at least part of suffering. Young goes on to argue—and I suggest we should treat these phenomenological reports with deference—that a suitable program of meditation can reduce resistance and therefore suffering.

12.4.3 Conclusion

In summary, there is considerable evidence that we could feel even intense pain without suffering very much from it. In such a state, we would be moved by pain to protect our bodies, but that sensation would not feel especially *bad*. Whether that badness can be *completely* eliminated, I am less confident about. For the relationship between pain and suffering doesn't seem like it should be entirely contingent—it is surely also an important fact that most moderately intense pains *do* hurt, and that it takes rather drastic or odd interventions to eliminate that hurt. I will return to this question and suggest tentatively there that although suffering can be reduced considerably, it probably cannot be eliminated without also eliminating important facts about agency and care (as in pain asymbolia).

Indeed, most of the interventions considered above are often considered to eliminate agency in their severe forms. That's most obvious with drugs. But the characteristic effect of hypnosis is often taken to be a diminished sense of agency (Barnier et al., 2008). Buddhist meditation practices are often, at least nominally, dedicated to the elimination of the individual self.

To conclude, I want to reiterate the distinction between three importantly distinct states that might be found in the clinical, experimental, and anecdotal literature. First, some interventions simply eliminate pain and *a fortiori* all ordinary pain-type motivational states. Second, some interventions might produce a state in which one feels pain but is entirely unmoved by it. If I am right, these interventions have to be drastic enough to compromise the agent's care for his body. Third and finally, some interventions might leave pain and the accompanying motivational force intact but vary the degree to which pain hurts. That there is *some* cognitive influence on suffering is, I take it, relatively uncontentious. The interesting question is then how far such interventions can go without compromising agency.

Before we get there, however, I want to consider the possibility of making the pain response *more* elaborate. Because I've treated suffering as a reaction to pain, it seems like it should be possible to have even higher-order reactions. I think this is possible and, further, that this helps resolve a long-standing puzzle about how to treat masochistic pleasures. Imperativism is not required to accept the account I will give, but I think that the two complement each other nicely.

13 Masochistic Pleasures

13.1 Introduction

Having dispatched the specter of pain asymbolia, I want to turn in the remaining chapters to the phenomenon of suffering and its relationship to the imperative that constitutes pain. Although the imperativist about pain needn't give a full account of painfulness, imperativism can shed light on why certain sensations *are* painful and why pains are paradigmatically so. To flesh out the account, I want to say more about the structure of painfulness and its relationship to both pains and other mental states.

To do so, I'll consider a long-standing puzzle for theories of pain: what should we say about masochistic pleasures?[1]

13.2 Masochistic Pleasures

13.2.1 The Cases

Most pains are simply unpleasant. Most pleasant experiences aren't painful. The neat division sometimes breaks down. Some people have painful experiences that they nevertheless find pleasant and find pleasant in part *because* the experience involves a certain kind of pain. Call these the *masochistic pleasures*.

Masochistic pleasures come in a variety of forms and range over several different categories of mental state. The paradigmatic ones involve bodily pains. Others involve emotional states. In his wonderful discussion of anger in the *Rhetoric*, Aristotle defines it as "an impulse, accompanied by pain, to conspicuous revenge for a conspicuous slight" (1378a31). Yet he also notes that anger is also "attended by a certain pleasure" (1378b1). When you dwell on anger, you also dwell on the revenge that you will get, and that's pretty great to think about. That is why anger is "sweeter by far than the honeycomb dripping with sweetness, and spreads through the hearts of men." Brooding anger has a similar structure to sensory masochistic pleasures: the emotion is painful, but by dwelling on it, one gets a certain kind of pleasure as well. Similarly, in his excellent paper, "Horror and Hedonic Ambivalence," Matt Strohl (2012) (continuing a theme from Hume) notes that many people like gruesome horror movies in part *because* they shock, scare, disgust, and horrify.

Here is a (non-exhaustive) typology of masochistic pleasures:

1 What follows is derived from Klein (2014).

LOOSE TOOTH Wiggling a loose tooth (getting a deep-tissue massage, pressing on a bruise, stroking a sunburn, etc.).

AGGRESSIVE VICTUALS Eating chilies (drinking strong whiskey, etc.).

HARDCORE SPORTS Running a marathon (powerlifting, etc.).

BODY MODIFICATION Getting a tattoo (piercing, etc.).

CLASSICAL MASOCHISM Being whipped (flogged, trampled, etc.) in a sexual context.

SLAP & TICKLE MASOCHISM Spanking (hair-pulling, biting, scratching, etc.) during sex.

ASEXUAL MASOCHISM Being whipped (beaten, trampled, etc.) in a *non*sexual context.

OBSESSIVE ANGER Dwelling on one's anger (jealousy, heartbreak, etc.).

AESTHETIC AMBIVALENCE Enjoying scary (gory, tragic, etc.) works of art.

This list is only a first pass and includes some cases that I will later distinguish. You might balk at lumping all of these together. Part of my job will be to convince you that they are different species of the same phenomenon. That said, the list is a useful first pass. In each case, we find people engaged in an activity that they sincerely claim is painful, and yet they also claim to find it pleasant. Further, there is good reason to think both avowals are true. The activity is unquestionably painful for people who *don't* find it pleasant. That's easy enough to confirm. Yet people don't just *claim* to find that painful feeling pleasant—they also pursue it. That is usually good evidence that someone is being sincere about finding something pleasant. So we have good evidence that the masochistic pleasures are real.

Masochistic pleasures present several puzzles. They are at best rare: most sensations are either pleasant or painful (or neither). Few are both. So the unusual nature of masochistic pleasures makes them worthy of attention. Further, one might think that this rarity is not just a matter of chance but a deep philosophical fact. On many theories, pleasantness and painfulness positively exclude one another. Pleasures are good; we like them. Pains are bad; we dislike them. The particular theses vary, but many theories assume that some such pair is true, and so that pleasantness and painfulness are by their very nature fundamentally opposed. Yet there are the masochistic pleasures. What do they show?

Imperativism might seem to be at odds with the fact of masochistic pleasures. There is a superficial problem here: masochistic pleasures seem to involve motivation *not* to protect yourself from the painful thing you're doing—in fact, you seem to be motivated to continue. This needn't concern the imperativist. The masochist being whipped is still motivated in virtue of the pain to protect himself, although there might be other facts that prevent him from *acting* on that motivation (including the fact that he takes pleasure in the pain of being whipped). So there is no more problem with the imperatives involved than there is in an ordinary case where one overrides the primary motivational force of pain.

Nevertheless, there is an interesting interaction among pain, the painful, and the pleasant in masochistic pleasures. Chapter 4 argued that painfulness and pleasantness might admit of a certain kind of recursive structure. That is, painfulness can take mental states other than pains as its object and can in turn be taken as an object. One possibility, then, is that painfulness and pleasantness might take *each other* as objects.

I think something like this is the key to understanding masochistic pleasures. Masochistic pleasures are often thought to be one of the distinctive puzzles for a theory of pain. I think by carefully separating pain and painfulness, we can shed light on this puzzling phenomenon. Further, masochistic pleasures offer an example of a case where there are plausible reasons for finding a certain sensation pleasant or painful, and so shed light on the sort of appraisals that might constitute such judgments.

13.2.2 Motivating the Problem

Many authors have argued that there is no *real* problem of masochistic pleasures. I want to canvass a few arguments along those lines and show why they are inadequate. That will also sharpen up the problem and clarify the parameters of an acceptable solution. Note that many of these debunking explanations do account for *some* putative cases of masochistic pleasures. My claim is just that they don't capture all of them.

Masochistic pleasures are sometimes presented as if they involve a normally painful *stimulus* that someone happens to find pleasant. Call this a "crossed-wires" account of masochism. The crossed-wires account may (for all I know) be true of certain people—physiology contains endless mysteries. In contrast, most masochistic pleasures involve something that actually *hurts* and is described as such. As George Pitcher (1970) puts it,

> [The masochist] does not find the pain inflicted by the whipstrokes of his sexual partner pleasant. What an absurd idea! If he did find them pleasant, he would look for a more aggressive mate. No, the pain is fearfully unpleasant, and that is precisely why it excites him, precisely why he wants it. Without the suffering, the whole exercise is pointless. (485)

I will therefore restrict the discussion to experiences that involve, and are described as involving, some degree of suffering.

If what is pleasant and what is painful are wholly distinct states, then we also don't have a puzzle. Call these *debunking* explanations. Debunking explanations are most common with SEXUAL MASOCHISM: the line goes, crudely, what's painful is the whipping, what's pleasant is the sex, and there is no mystery in either case. A debunking explanation can take a variety of forms. Sometimes pain is merely *endured* for the sake of some pleasant thing. These go back at least to Cicero, who claimed that,

> Nor again is there anyone who loves or pursues or desires to obtain pain of itself, because it is pain, but occasionally circumstances occur in which toil and pain can procure him some great pleasure. To take a trivial example, which of us ever undertakes laborious physical exercise, except to obtain some advantage from it?[2]

When Mucius Scaevola thrust his hand in the flame to intimidate the Etruscans, he felt pain. He put up with that pain for the sake of glory, which was something that he found pleasant. What he did wasn't pleasant: in fact, his demonstration was effective precisely *because* he was willing to do something so awful.[3] On a milder note, I suspect that some—but not all—gym-goers and tattoo-acquirers merely endure pain for the sake of something else that they desire.

Similarly, pain can be *used* to create some other, pleasant effect in virtue of some of its physiological properties. Some painful things release endorphins,

2 *De Finibus*, Book 1, sections 323, Rackham translation. The original Latin is best known as the source text for the "Lorem Ipsum" typesetting aid.

3 The story is recounted in Book 2 of Livy's *History of Rome*.

and endorphins make you feel mildly euphoric. That is a pleasant sensation—one, however, that's distinct from the pains that caused it. There is also a non-specific quantity of "arousal" that seems to be partly transferrable between sensations. In a well-known study by Dutton and Aron (1974), for example, subjects crossing a high, rickety bridge were more likely to show evidence of sexual attraction to a female experimenter than were subjects crossing a low, stable bridge nearby. One explanation for this effect is that the arousal from fear is partly transferrable over to sexual desire.[4] Some apparent masochistic pleasures in vanilla sex life can probably be put down to this phenomenon.

Each of these debunking explanations captures a few of our cases. Many remain. LOOSE TOOTH-style cases in particular don't seem to be accounted for: there, the pain seems to be precisely what is sought. Further, in cases where pain is merely endured or used, the subject is usually quite open to getting the goal without the pain. Whereas in cases of masochistic pleasures, people actually seek out the painful experience as such and often seem to think that it would be diminished without the pain. This is most obvious in the case of SEXUAL MASOCHISM. Some people pay sex workers to whip them; this is a specialized service, and so costs more than simply paying for sex would. In general, I doubt any theory that posits two wholly distinct objects is going to handle most of our cases. That's because masochistic pleasures are pleasant *in virtue of* the suffering not merely *despite it*. The connection between pain and pleasure appears to be tighter, and that's precisely what generates the problem.

A further category of explanations suggests that pleasure and painfulness qualify two different states of affairs that are not wholly distinct. Call these *contextual* explanations. In each case, one appeals to a larger context in which the pain is embedded and argues that what is pleasant is the larger context. One might claim, for example, that the pain is a necessary *constituent* of a context that is pleasant (Pitcher, 1970). The marathon runner feels pain. Feeling pain is a necessary part of running a marathon, and he finds running a marathon pleasant. Contexts in this sense are simply facts about the world. It is no surprise that contexts might have a compositional structure, and so a pleasant context might be partially constituted by a painful one. Masochistic pleasures, then, just are ones where that composite structure obtains.

4 The authors link the explanation of this effect to Schachter and Singer (1962); I think Barrett's (2006) constructivist approach would produce a similar result.

Unlike debunking accounts, the contextual account has a story about why the marathon runner doesn't just take a cab: the thing he finds pleasant is partially constituted by something he finds painful, and so the painful part can't be subtracted out. However, this simple contextualist story doesn't seem promising. LOOSE TOOTH-style cases don't seem to involve any further end. I often eat hot chilies on my own because I like the burn. ASEXUAL MASOCHISM appears to exist more or less for its own sake. Of course, one can always hypothesize a larger context: one might claim that the overarching goal in the case of loose teeth is the pleasure of adult teeth, or the end of pain, or whatever. Bearn (2013) plausibly notes that these jury-rigged contexts would overgeneralize absent further constraints. I might like watching a movie at my favorite arty theatre even though the seats are uncomfortable. Painful seats are part of the grungy aesthetic, so I wouldn't want them gone. There I have a pleasant experience that has a painful experience as an ineliminable constituent, but I'm not feeling a masochistic pleasure.

Fred Feldman (2004) suggests a more abstract contextual strategy, arguing that what masochists find pleasant is just the fact that they are feeling a painful sensation. *Why* they find this pleasant is left open by Feldman. But I take it that this is a kind of contextualist explanation: what is pleasant for masochists is *the fact that they feel some sensation to be painful*. That fact is necessarily constituted by, but distinct from, the facts that make the sensation painful.

I have sympathy for Feldman's account, and it is superficially similar to the one I will ultimately propose. However, I think Feldman's solution still overgeneralizes. I might awake from an operation and feel pleased that my leg hurts—the other option was amputation, say, so if I feel pain, I must still have a leg. But that is not to feel a masochistic pleasure. *Contra* Feldman, being pleased *that* I am having a painful sensation is not the same thing as taking pleasure *in* its painfulness. In general, it can't be sufficient for masochistic pleasure for a painful experience to be embedded in a pleasant experience (even necessarily or constitutively). That sort of embedding is relatively common. Most of our pleasures are admixed with some pains, and we often wouldn't trade those token experiences for other similar ones. But masochistic pleasures are relatively rare, more rare than such scenarios. An adequate theory should show both how masochistic pleasures can occur and also why they don't occur more often. Doing so will in turn go some way toward explaining the (incorrect) philosophical intuition that pleasure and pain exclude one another.

That said, I think contextual accounts are on the right track. Whereas they claim that what's pleasant is some composite *state of the world*, I will argue that it is a composite *sensation*.[5] Only in such cases, I claim, is painfulness *felt as* pleasant rather than merely being a part of some pleasing context.

13.2.3 The Plan

The goal of what follows will be to give an account of masochistic pleasures. Doing so requires answering two questions. First, one might ask *how* masochistic pleasures are possible. That is, one might ask for a philosophical story about pleasures and pains that accounts for masochistic pleasures. That might be a story that makes pleasures and pains compatible, or it might be a debunking explanation showing that the apparent incompatibility is misconceived. Second, one might ask just *what* is pleasant about the masochistic pleasures. This is an exercise in philosophically sensitive description. There is the class of masochistic pleasures. The goal is to say something about what they have in common and what differentiates them from other pleasures. That property, whatever it is, will be shared only at a relatively abstract level of description.

Most philosophers writing on masochistic pleasures have (understandably) focused on the how-question. I propose to begin with the what-question first. For one, it is an interesting question in its own right. For another, most answers to the how-question presuppose an answer to the what-question and quite often the wrong answer. Sorting out the what-question will thus constrain acceptable answers to the how-question. After telling a story about the what-question, I'll then step back and give an abstract characterization of masochistic pleasures that partly answers the how-question. I'll conclude with some general reflections about masochistic pleasure and its role in theorizing.

Before I begin, I want to make two important caveats. First, my target should be distinguished from *masochism*. Talking about *masochism* or *masochists* makes it sound like the important distinction is between *people*: that is, between folks who like pain and folks who don't. I think that is an unhelpful cut for a number of reasons. Most important, I think it is false: many people feel masochistic pleasures at least occasionally. Most of us have had a pleasantly

5 On the notion of composite sensations, see also Strohl (2012). I have qualms about Strohl's Aristotelian theory of pleasure and pain; briefly, I think there are numerous pleasures that don't arise from the optimal activation of a particular capacity. That said, I think the overall structure of his account is similar to mine and appealing.

painful deep tissue massage or felt the thrill of water that's just shy of unbearably hot or cold. Conversely, even people who self-identify as masochists don't like all kinds of pain—typically they like only specific sorts of pain in specific sorts of circumstances (Stoller, 1991). So what masochists like is *kinds* of pain not pain as such. In what follows I'll occasionally use the term "masochist," but this should be understood as shorthand for a "person who self-identifies as a sexual masochist." That is a flexible category, but it does let me interface more easily with the literature on sexual masochism. Some, but not all, masochists in this sense feel masochistic pleasures—a nontrivial proportion engage in their practice for other reasons.

Second, and along the same lines, I'm going to assume that most people have felt at least some masochistic pleasures at some point in their lives. The first category on the list seems to be widely attested: when I give talks on pain, I often get questions about the childhood pleasures of wiggling loose teeth and the adult pleasures of the deep-tissue massage. If masochistic pleasures are in fact widespread, then we should avoid explanations that treat them as manifestations of a pathology. Along related lines, there is no reason to think that the masochistic pleasures are especially associated with sex.[6] LOOSE TOOTH, AGGRESSIVE VICTUALS, and HARDCORE SPORTS certainly don't seem sexual in nature. Further, some self-identified asexuals enjoy various forms of traditional masochistic practices such as whipping and spanking.

13.3 A Penumbral Theory of Masochistic Pleasure

13.3.1 Motivation

Here's the rough idea: each masochistic pleasure involves some sensation that is painful. What is pleasant is *the painfulness itself*. Because painfulness is rarely pleasant, we need an explanation of what is special about these cases. I think the answer lies in particular features of token painful experiences.

As a warm-up, consider the related phenomenon of *relief*. If you've been in severe pain for a while and it changes to a more tolerable level, then you feel relieved. What explains your relief? Facts about your pain: that it has *subsided*.

6 Absent some implausibly strong Freudianism, which claims that more or less everything is infused with sexual meaning. But even if you like that sort of story, masochistic pleasures need not be *especially* associated with sex, at least any more than the pleasures of cooking or model train building.

Relief comes, roughly, when you judge your pain is diminishing in intensity and is likely to keep doing so. This relies on a complex judgment that incorporates feelings about the pain's current intensity, memories about its past intensity, and beliefs about its likely future intensity.[7] You obviously don't feel relief in response to *all* pains. That's hardly a mystery. Not all pains have the right combination of features, and only those that do give rise to a feeling of relief.

I suggest that a similar story can be told for masochistic pleasures. What is found pleasant is a particular *quality* of the unpleasant sensations involved: that is, some feature of the sensation that distinguishes masochistic pleasures from the others. The goal of this section will be to spell out that quality.

13.3.2 Edges

For the strategy to work, I need some quality that might be plausibly felt as pleasant, and one that is rare enough that many instances of unpleasant sensations don't involve it. I have a candidate that I think is largely overlooked in philosophical accounts of masochistic pleasures, although it is quite common in first-person accounts of sexual masochism.

To begin, consider LOOSE TOOTH cases. Wiggling a loose tooth is painful. It is not *so* painful, however. It could be worse. If it got bad enough, then you'd stop. Similarly with pressing bruises, stretching muscles, and so on. Further, the painfulness you feel in such cases isn't at some arbitrary level: typically, it is right on the edge of what you can bear.

That is why one finds a certain *fascination* at work in these cases. People don't push a loose tooth once and then stop. They keep returning, pushing right to the edge of what they can bear, and then backing off, sometimes going over, and in general making exploratory sallies right around the borderline where the sensation becomes too much. The repetition is not just because the sensation is pleasant (not all pleasant sensations produce this fascination). Chalking this up to mere pathological compulsion isn't plausible in many cases. Rather, I think the fascination is a product of the more general process of finding the edge of bearability.

That same structure—flirting around the edge of something that is nearly too much to bear—comes out in other masochistic pleasures as well. In AGGRESSIVE VICTUALS, the pleasant sorts of "burn" are those that are intense but not

7 For a useful discussion of these factors, as well as for some practical applications, see Redelmeier et al. (2003).

quite too intense. Mere whiskey doesn't do it, nor merely spicy food—it has to be the high-proof or Sichuan stuff, respectively. Similarly so with HARDCORE SPORTS. I think there is a reason why you get reports of masochistic pleasure from marathon runners rather than, say, pickup basketball players. Most pain felt during exercise is merely painful; when there is pleasure in the pain, it comes from really pushing your limits.

Finally, there is certainly this component in a lot of sexual masochism. Masochists frequently discuss the process of "finding their edge" and "pushing their limits." Narratives of masochism talk about pain pushed so that they can hardly bear it or of going just over the edge and pulling back over and over. In a popular manual on BDSM, the authors actually give instructions for how to appreciate masochistic pleasure if you're unfamiliar with it:

> Find a way to give yourself a stinging or thudding sensation, one that doesn't damage your body. We want to focus on sensation, not injury... Deliver each stroke at your Resilient Edge of Resistance, right at the place where the pain is enough to make you gasp, but not so intense that you withdraw from it completely. (Taormino, 2012, 496)

Here the point is quite clear: masochistic pleasure is found right on the edge of what you can take.

Indeed, there is often an explicit connection drawn between masochism and other edgy sorts of endeavors. Some sexual masochists draw a connection between enduring pain in sports contexts and sexual masochism. In her controversial memoir, ballet dancer and critic Toni Bentley (2006) says of the masochistic pleasure she takes in anal sex that,

> It answers the call of my physical masochism. It re-creates the physical extremism of dancing, the discipline, the striving for perfection. It is my being in extremis. (144)

Here we have a direct link between sexual masochism and the sorts of edge-pushing discipline one finds in other bodily pursuits. Many authors have noted that some masochists appear to get pleasure from humiliation rather than bodily pain (Baumeister, 1989). Humiliation is unpleasant, and like all unpleasant sensations, it comes in degrees. I suspect the process of humiliating a masochist involves similarly finding that edge where humiliation is just shy of unbearable. Discussions of sexual masochism often stress the importance of conversations beforehand talking about the masochists' limits, what they'd like to do, and so forth. This conversation plays a crucial role in establishing consent. But I suspect it also does something else: it lets the top know where those limits are precisely because playing with limits is part of the point.

I think this structure also handles the emotional varieties of masochistic pleasure. Jealous brooding over the whereabouts of an ex-lover has a similar quality to wiggling a loose tooth. Robert Hass' wonderful poem "Faint Music" details the jealous musings of a desperate man left by his lover. After laying out an intricate, painful scene wherein the narrator envisions his ex with a new lover, Hass says, "And he, he would play that scene// once only, once and a half..."—an admission obviously false, as he has clearly run through this many times, and will again, suggesting a similar dwelling on pain. To return to Aristotle, worrying oneself over a slight and the rage it provokes has a similar structure: the combined feeling is one of being pushed away from one's current state (caused by the slight) and toward revenge (as a solution). Too much either way would be either demoralizing or energizing—it is only on that edge that one keeps going back to the slight and worrying it.

In summary, what appears to be common to, and probably distinctive of, cases of masochistic pleasure as such is this process of pushing a painful feeling just to the limits of unendurability. Call this the *penumbral* theory of masochism. Masochistic pleasure is possible in cases where pains are in the shadowy border just shy of being too much to bear. So we can elaborate the structure of masochistic pleasures a bit further: the masochist is in pain (or some other negative experience). Pain is painful. That pain is almost, but not quite, too much to bear. Having a pain that is almost, but not quite, too much to bear is, under the right circumstances, pleasant. Hence, the masochist finds something painful, and finds that painfulness pleasant. The suffering and pain come about for different *reasons*: suffering because there is is a pain, pleasant because it is a suffering with the right combination of qualities.

13.3.3 The Penumbral Theory

I suggest that it is this feature—the penumbral quality—of pains that is found pleasant in cases of masochistic pleasure. I'll call a "penumbral sensation" any token sensation (or other mental state) that is resting on that edge of unbearability.

Now, it would be nice for my account if all instances of flirting around the edge of unbearability were pleasant. That's not the case. I might, for example, soldier on at a grim historical museum, doling out the horrors just enough to advance my education. I might even be quite aware that I'm on the edge of what I can take. This would be simply unpleasant, despite being near the edge of bearability. (Not necessarily: I might, like Leontius, find some grim fascination in the corpses. But that is surely not necessary, and I take it that such a trip might be just unpleasant.)

So there is a little more explaining to do about *why* and *when* some penumbral sensation is pleasant. The penumbral theory says that it is the awareness of being on the edge of unbearability that is pleasant. *That* aspect is tracked by pleasantness. It is found pleasant for some particular reason. That reason may not be accessible to the person who feels masochistic pleasure, but I think we can reconstruct several possibilities from those who are more reflective. Note that I am a pluralist about these reasons: there are many possibilities for *why* someone finds penumbral sensations pleasant. That said, I think the available reasons divide into a few broad (and nonexclusive) categories.

First, one can find being on a borderline pleasant because it is *novel* and novelty has a certain pleasure to it. This may be partly what is going on with children and loose teeth. Children have to learn to control their reactions to pain; finding the edge of unbearability might be pleasant precisely because it is a surprising and novel discovery. This novelty may wear off for most of us, although maybe not all. Many authors have remarked on the sheer variety and inventiveness of masochistic sexual practices. This suggests that some of the pleasure involved there may be traced back to a similar appreciation of novelty: keeping someone on the edge of bearability might be pleasant because it is done in a *new way* rather than the tired old ones.

Second, and I think probably commonly, there is a pleasure that comes from exercising self-control. That comes out clearly in the Bentley quote and, I think, in many sports cases as well. By *deliberately* pushing up to the edge of bearability, one is also exercising the control necessary to stay on that edge. Your capacity for self-control in the particular case probably has to be something that you value about yourself. If you do, then it is not surprising that you might take pleasure in exercising such control by being on that edge. It also shows why you need to be close to the edge for it to be pleasant. If the pain were too little, then control wouldn't be necessary; if it were too much, then control would be overwhelmed.

Third, in the case of sexual masochism (both classic and slap-and-tickle sorts), I think that penumbral sensations can be pleasant for a variety of reasons. A lot of erotic life lies just on the edge of being swept away by uncontrollable feelings. So it's not a terrible surprise that standing on the edge of such a feeling (even a painful one) can be sexy. Further, penumbral sensations can be pleasant because they stand as unique signifiers of trust and intimacy. You are around the edge of what you can bear, and someone is making it so, and you trust them to keep it thus and no more. In refusing to pull back, in staying right on that edge, you're having a sensation that is born of a combination of desire (in the sexual sense, not the philosophical sense) and deep trust. Again, when you put it this way, it's not surprising that penumbral sensations can end up erotic.

I think the second and third reasons often combine. Lovers often prolong their pleasure, holding off on getting to the good bits. That results in a certain amount of pain from temporarily frustrated desire, but that sort of pain has a sweetness to it. The mixture of self-control and trust is, again unsurprisingly, quite pleasurable. Of course, some cases of sexual masochism may emphasize one over the other. I suspect, however, that what seem like pathological manifestations of sexual masochism do not involve a pathology of *sex* so much as a pathologically exaggerated sense of self-control or submission.[8]

Fourth, and finally, in some of these cases, pushing your own boundaries helps you grow and change, and that feeling of growth and change is what is perceived as pleasant. Each time you replay a scene in your mind, one you can't control externally, you're pushed to control internally. Perhaps that is precisely the space where important sorts of narrative work happen. By reflecting on things at the edge of bearability, you slowly learn how to bear them. So what is pleasant there is the feeling of growing beyond patterns that you're trapped in: you're finding out that a slight or a loss that seemed unendurable is perhaps bearable after all.[9]

13.4 Contextualism *Redux*

It is worth saying something about how my account differs from (and is preferable to) a contextualist theory. The contextualist, recall, said that what was pleasant was an overall context in which a painful experience was a constituent. The contextualist might appeal to the reasons I just cited to characterize what's pleasant about the overall context. What's pleasant, they might claim, is not the penumbral sensation but the overall context in which it occurs. That context is constituted by the reasons to which I just appealed.

8 Note that the DSM–5 no longer recognizes "atypical sexual interests" as pathological per se. To be diagnosed with a paraphilic disorder, you must either feel "personal distress" about your desires or desire things that are morally and legally problematic (American Psychiatric Association, 2013).

9 See also Mollena Williams' contributions to Taormino (2012), which discusses this effect in the context of sexual masochism.

In response, I suggest that this contextualist solution runs together the *object* of pleasure with the *reason why* we find that object pleasant. These should be kept distinct. I have a statue of a camel that I find particularly pleasing. There are many reasons why: its realistically defiant pose, the aesthetic harmony of the *sancai* glazing, and so on. Yet *what* I find pleasing is simply the statue. The point is not merely that I can appraise my statue without these reasons coming to mind (although that is also important). Rather, to say I find the reasons pleasant is to over-intellectualize my experience. What I like is the statue not my reasons. So too with masochistic pleasures. What is pleasant about wiggling a loose tooth is not novelty. It is the penumbral sensation, which is found pleasant *because* it is novel.

Because the penumbral theory separates the object of pleasure and the reasons for pleasure, it succeeds where contextualism faltered. On the one hand, we saw that contextualism had difficulty coming up with an appropriate larger context in cases such as LOOSE TOOTH. The penumbral theory faces no such difficulty: penumbral sensations are pleasant, not some larger context. These are always present in cases of masochistic pleasure, even when they don't constitute some further pleasing situation. Conversely, contextualism overgeneralizes: there are situations constituted by painful experiences and that we find pleasant, but that aren't instances of masochistic pleasure. The penumbral theory avoids this. Merely finding a larger context pleasant is not sufficient for masochistic pleasure. Feeling masochistic pleasure requires that you find the penumbral sensation *itself* pleasant. That is not assured, even if the sensation is an essential component of a larger context.

13.5 A Structural Account of Masochism

The penumbral account is meant to say what masochistic pleasures are. Suppose you're convinced. Then we can abstract away from the account a bit to say something about how masochistic pleasures are possible. Doing so will also add to the defense of the penumbral account, for those who aren't quite convinced yet. Abstracted away to a structural core, the thesis is as follows. Masochistic pleasure occurs when:

1. There is a first-order state (or states) such as a bodily pain, which is
2. Felt as *painful*, and that quality of painfulness is further
3. Felt as *pleasant*.

To make this more concrete, suppose someone takes pleasure in being spanked. The current account says that there will be three distinct features that jointly characterize the experience. There is a first-order sensory state, the bodily pain, that arises from the spanking. That sensory state is *painful*. It hurts. That quality of painfulness is felt as pleasant. The distinctive contribution of the penumbral account is to explain just when and why painfulness can be pleasant. As I've argued, it depends on a particular feature of the painfulness—namely, its penumbral quality—that is found pleasant for one of several reasons.

On this version of the story, what is pleasant is the painfulness of the first-order sensation rather than the first-order sensation itself. I think this is preferable to a story on which both qualify the first-order negative state. It allows us to distinguish a related but distinct phenomena, what we might call *bittersweet pleasures*. Bittersweet pleasures involve situations that are mostly pleasant, but that very pleasure can be the cause of a kind of pain. I might, for example, feel pleasure watching my child graduate high school—but with that, a certain kind of pain because I know that this pleasure will fade and diminish when he moves away from home. Masochistic pleasures involve a primary base of painfulness, with a certain attenuated pleasure on top; bittersweet pleasures involve the reverse. If pleasantness and painfulness both qualified the first-order sensation, then masochistic and bittersweet pleasures would be indistinguishable, but they seem to be phenomenologically distinct.

Masochistic pleasures both draw you on (because they're pleasant) and push you away (because they're unpleasant). Pushing you away and pulling you toward are both distinct causal roles. But now we can see why there is nothing problematic in principle about both going on at once. The causal relationships in virtue of which something is painful and in virtue of which it is pleasant can be exemplified in the same composite sensation. There is no more a problem here, as far as the causal story goes, than there is in a single physical object both pulling you and pushing you at the same time. In most cases (as with physical causation), one expects that the pleasant and painful will be such for *different reasons*. That is, the pushing and the pulling depend on different aspects of the sensation, and so obtain for distinct reasons. That is precisely what is going on in the case of masochistic pleasures.

Summing up, here's the proposal: pleasantness and painfulness are feelings. They are feelings that take as their objects a wide variety of different mental states. Painfulness and pleasantness are therefore *higher-order* mental states. Further, as higher-order mental states, they seem to track some facts about the lower-order states to which they attach. It is not surprising that bodily pains are felt as painful, so steps one and two of the structural story are satisfied relatively easily. The penumbral account comes in at step 3: it explains why painfulness *of a certain kind* can be felt as pleasant. With that, we have a story about masochistic pleasures.

Masochistic pleasures seem odd because it is hard to imagine how one and the same sensation could be painful and pleasant. By separating out sensations (even pain) from qualities of painfulness and pleasantness, however, we find that there is nothing in the structure of either of those qualities that excludes the other. There are good reasons why they rarely coexist, of course, but it is not impossible. I suggested that the place where we often find them together is in the shadowy, penumbral area where a painful sensation is nearly unbearable. With the right sorts of motivations, however, one can find being on that edge pleasant. Masochistic pleasure is not philosophically perplexing; in the case of sexual masochism, it may not even be surprising.

14 Imperatives and Suffering

14.1 A Return to Suffering

Throughout the book, I have distinguished between pain and suffering (or *hurt*, or *painfulness*). Suffering, I suggested, was best understood as an attitude taken toward pain. That attitude could also be taken toward a variety of other sensations. Hence, it was possible to be painfully hungry, painfully tired, or painfully lonely—this even though none of those sensations involved pain in the sense I have been discussing.

I have emphasized throughout that suffering is also a motivating state, but one that motivates different ends than pain itself. As Bennett Helm (2002) suggests:

> Although some pain behavior is mere reflex, as when you jerk your hand away from a hot stove, in general pain motivates intentional action. Thus, if your arm is strapped down as I bring a lit candle to your hand, you may try to blow the flame out or push me away, actions that are rationally motivated precisely because of the badness of the pain you feel.

I suggested that pain motivates you in a way more complex than mere reflexes would allow; in Helm's case, the pain of the lit candle motivates you to protect your arm by getting it away from the candle (of course, in the imagined case, you can't, but that doesn't mean that you're not motivated to do so). But because the pain *hurts*, that gives you additional reason to act—to blow out the candle or more generally remove the source of the sensation—which is exactly as Helm describes.

The focus of this book has been pain, not suffering. As far as suffering goes, the primary explanatory goal for the imperativist is to say why imperatives should produce this intense reaction. Pains are the paradigmatic painful sensation; ideally, there should be some especially tight connection between pains and suffering. In the course of arguing for the distinction between pain and hurt, I suggested that some mild pains don't even hurt, and that even severe pains might not be painful. That makes the explanatory task more pressing: given that mild imperatives need not hurt, why *do* strong pains mostly hurt?

I don't think I have a complete answer to this question. I do have a few suggestions. Sensations—and pains in particular—hurt because they threaten our agency in a particularly deep and intimate way. First, I'll reflect on what suffering might be. Then, I'll talk about *why* pains are especially poised to make us suffer.

14.2 Two Perspectives on Suffering

14.2.1 Theorizing About Suffering

Although I have no theory of what suffering is, I do have various things to say about it. It is a motivational state. That motivation is, I've argued, directed toward other mental states. (It may also be directed at the world in some cases; I remain neutral on that.) That makes suffering a higher-order mental state: it is a mental state that is partly directed toward some other mental state. In chapter 13, I relied on the fact that this recursive structure is shared with pleasure to show how various higher-order states combined can make sense of masochistic pleasures.

Whatever suffering is, it appears to be fairly strongly encapsulated. It's hard to talk yourself out of suffering from pain. Suffering also has a distinctive phenomenology: one reason I carved off suffering from pain, remember, was precisely because there seemed to be a phenomenology common to pains and other painful states.

Finally, suffering is a negatively valenced evaluative state: following a suggestion of Aaron Smuts (2011), we might say that hurt (whatever else it is) is identical to the way that certain states feel *bad*. This is something that strikes us, immediately, about the experience of states that feel bad. As Baier (1962) puts it,

> To say that we are finding something painful is not to misstate the case. It really is a finding, in that it is quite unargued and unarguable. One is not summoning up expertise, marshaling pros and cons, citing evidence or precedents, and so forth. On the contrary, the painfulness hits one with inescapable force. (5)

It is because we suffer from our pains that we dislike them, that we are motivated to get rid of them, and so on. But that is secondary to what pains *are*, which is imperatives to protect a particular body part.

While I will not give a theory of suffering, I do think that making the distinction between pain and hurt is a useful step toward giving such a theory. Indeed, I think there are promising candidates in the philosophical literature. I will briefly discuss two and suggest how they might actually be taken to converge.

14.2.2 Evaluativism

The first possibility is what I've called *evaluativist* theories. Evaluativist theories treat pains, in Helm's (2002) terms, as "felt evaluations" of the badness of a particular state. In addition to Helm, David Bain has also offered an evaluativist account of pain, and Korsgaard's (1996) discussion of pain has excellent reflections on painfulness that strike me as evaluativist.

Bain, Helm, and Korsgaard present their views as accounts of *pain*. In chapter 10, I argued that evaluativism is the wrong theory of pain. Evaluation is a normatively heavy notion, too much so for a basic state such as pain. Further, the states that are motivated by taking something as bad aren't necessarily the biologically adaptive ones that pains motivate. However, I think evaluativism is a plausible theory of *suffering*. To suffer is to feel one of your own sensations as bad. That gives you reason to end that sensation and can thereby motivate a variety of different actions toward one's pain (or hunger, thirst, or whatever).

Indeed, I think imperativism has a lot to offer evaluativism, now considered as a theory of suffering. One problem that has always plagued evaluativist theories of pain is the question of why we can't just *stop* disliking pains. That's an especially pressing question for maladaptive pains such as phantom limbs: *why* continue to dislike the pain so vividly, given that you know it doesn't signify damage, and that disliking it constitutes suffering, which is bad? The considerations in chapter 12 offer the beginnings of an answer: pain feels bad because it impinges on our agency in a particularly direct way. It's difficult to imagine circumstances under which it might cease to be bad. Most of the cases I discussed in chapter 12 are clearly pathological—the ones that aren't involve strange changes to the experience of agency.

Alternatively, an evaluativist about suffering might take up a suggestion made by Immanuel Kant (1912) in the *Critique of Judgment*. In §10, Kant writes,

> Consciousness of a presentation's causality directed at the subject's state so as to keep him in that state, may here designate generally what we call pleasure; whereas pain is that presentation which contains the basis that determines [the subject to change] the state [consisting] of [certain] presentations into their own opposite (i.e. to keep them away or remove them).

Unpacking this a bit, Kant treats pleasure and painfulness as a kind of higher-order appreciation of first-order sensations. In particular, it's an appreciation of the effect that those sensations have *on* us. As Rachel Zuckert (2002) puts it,

> Kant does not mean by "feeling" that pleasure is an indefinable, primitive mental state (as, for example, empiricists tend to assume). Instead, pleasure is apparently a mental state that is 'about' another mental sate and, specifically, about the continuation in time of that mental state. For example, pleasure in eating chocolate on this account would be the 'feeling' or 'consciousness' of wanting to continue sensing (tasting) the chocolate (or, more precisely, the feeling that the presentation of chocolate is 'causing' one to want to stay in the state of having that presentation). (240)

Thus, we have painfulness and pleasantness as higher-order mental states: they are directed at other mental states, and they are judgments that those states bear a certain motivational relationship to us. Note that what's shared among these sensations is relatively abstract: their relationship to you rather than any of their first-order nonrelational properties. That's good for my imperative account because the *contents* of pain, cramps, itches, and so on are all quite different. Thus, painfulness is an appreciation of a quality that the object of pleasantness—here conceived of as a mental state—has. It is a higher-order judgment *about* the sensation and its own motivational force. One attraction of the Kantian version might be that it does not require judgment about the first-order states as good or bad, and so is conceptually a bit lighter. The downside might be that it is less clear why these self-undermining states would be felt as bad, although again considerations about the relationship between them and one's sense of agency might fill the gap. Another downside might be that the Kantian version applies most cleanly to imperative sensations, but it's not obvious that all painful states involve first-order imperative sensations (this is surely not the case for the *pleasant*, which is another reason to worry). Sufficient cleverness about the judgment involved might overcome these problems.

14.2.3 Second-Order Imperativism

A second possibility, and one that I'm rather fond of, is that suffering is constituted by an imperative. This would be a *second-order* imperative directed toward the first-order sensation. Second-order imperatives are possible, although relatively rare, in ordinary language; consider "Don't do as she says!" or "Follow these instructions to the letter!"

The second-order imperative in the case of suffering might simply be, "Don't have *that* sensation!" Manolo Martínez (2010) suggests an account like this, which gives the content of pain as, "Don't have this disturbance!" (76).

I suggested reasons to reject this as an account of *pain* in chapter 5. But it seems to me to be quite plausible as an account of *suffering*. Satisfying the second-order imperative would involve taking steps to eliminate the sensation: in the case of imperative sensations, it might involve satisfying the imperative or taking steps (such as swallowing an ibuprofen) to eliminate the imperative. Second-order imperatives can also be general in scope: any sensation could be referenced by the included demonstrative element.

Second-order imperativism is flexible enough to include babies, animals, and other cognitively limited creatures. The general form of the second-order imperative can be shared between them and us: when I suffer and when a dog suffers from pain, we both have a first-order imperative and a second-order imperative directed at the elimination of the first-order one. The capacities I can use to satisfy the second-order imperative are more flexible than those available to the dog, however (I know where the ibuprofen is, and he doesn't). Cognitive capacities can enrich the satisfaction of second-order imperatives without changing their fundamental nature.

The primary disadvantage to second-order imperativism is that the second-order imperatives are themselves a bit mysterious. *Why* have a state that drives you to get rid of pain? After all, you might think, the pain motivates you to get rid of pain in the *best* possible way, namely, by protecting yourself. That might be a relatively slow route to its elimination, but quicker routes (such as ibuprofen) run the risk of encouraging more damage.

It is here that a hybrid account might get some purchase. The reason why you should get rid of pain is because it *is bad*. It's bad for the reason that evaluativism has latched onto: pains of even moderate intensity systematically work to thwart your agency and limit what you can do. Second-order imperativism might be able to accommodate this insight. I argued in chapters 6 and 7 that imperatives need an issuer. The issuer of second-order imperatives is unlikely to be the body. Rather, I suggest, second-order imperatives might be issued by the *agent*. That is, suffering is a second-order command that *we* issue to *ourselves* because we have judged a first-order sensation to be bad in some way. Agential-level commands, in this sense, would still have to be relatively quick, encapsulated, involuntary sorts of states—for again, we can't talk ourselves out of suffering. Note too that this wouldn't be a pure form of imperativism: suffering would consist of both an evaluation of a first-order state and a command with respect to that state.

The hybrid picture would read something like this: Pain is a command issued by our bodies that serves to protect our integrity as physical things. Painful*ness* is a command issued with respect to pain (and many other states) that serves to protect our integrity as *agents*. For overwhelming pain can, by its very nature, destroy us as persons. Suffering would then be at once a recognition of this fact (an evaluation) and an exhortation to solve it if you can (an imperative).

14.3 Some Reasons Why Pains Hurt

The preceding sections discussed a few speculations about what suffering amounts to. There is another, distinct question about why we suffer in the first place. What follows is, again, speculative. However, I think there are several plausible things the imperativist can say. Further, I believe that some of them are distinctive to imperativism, which is another point in its favor. I'll consider a few general reasons and then move to the ones that seem to be more intimately associated with the imperative nature of pain.

14.3.1 Hating the Messenger

A first reason why we dislike pain is simply because it keeps bad company. As Hall (1989) argues, "Why do we find pains so unpleasant? Because they accompany nociceptual reports of bodily damage, and bodily damage is something we don't like" (647). This badness needn't have anything to do with pain itself; as Hall suggests, "It is like the ruler who slew the messenger who brought the bad news; pain sensations are no more inherently bad than that messenger" (647).

I deny, of course, that pain is a nociceptual *report* of bodily damage, but Hall's claim seems plausible enough. Often our pains are associated with things we dislike. Bodily damage is usually no fun. Nor are fear, anxiety, and so on. So pains can be disliked just because they are regularly associated with things that we dislike.

The sense of "dislike" here should, I think, be considered as involving *associational* processes rather than personal-level doxastic ones. The simple association between pain and other bad states is enough, over time, to make us dislike them. Associational processes do not require every pain to be associated with bad things or even very many of them. An ill-spent night in my youth with a bottle of schnapps was sufficient for me to dislike the smell of artificial cinnamon for the following twenty years. I know full well that most things that smell like cinnamon won't make me violently ill; nevertheless, I still can't stand the smell. Similarly so with pains. I might know that this particular pain isn't associated with something bad, but pain has a terrible track record, and I can't help but dislike it.

These associative processes can both diminish and enhance the badness of pain. On the one hand, there is a well-known phenomenon where children take their cues about the badness of pain from their parents. As Hall (1989) writes,

> who hasn't seen a child fall down or in some other way "hurt" itself (that is, do minor bodily damage to itself), and then look to its parents to see how it should react. If the parents are smart and show no concern, the kid picks itself up and goes on its merry way with no further fuss; but if the parents indicate that something bad has happened to the child, it will react as if it is feeling something bad. (653–4)

Conversely, chronic pain syndromes often involve a negative feedback loop in which fear of increased pain from movement is associated with present pain, which makes that pain more unpleasant, which in turn increases fear of pain via movement, and so on (Vlaeyen and Linton, 2000).

Finally, as numerous authors have suggested, some of this dislike could be innate.[1] It is not merely that pain is associated with bad things in our lifetimes, but that it has been consistently associated with bad things across the course of evolutionary history. An innate component would explain why even children seem to think there's *something* bad about pain, and why it's hard to de-associate pain and badness.

14.3.2 Pain and Frustrated Plans

The previous story, while part of the badness of pain, is a rather general one; it can apply to any sensation that one ends up finding painfully averse. The imperative sensations may also be disliked for a more specific reason.

Each imperative sensation demands a certain type of action. Typically, that action won't be the action you're otherwise inclined to do. Hence, imperative sensations frustrate your ordinary plans in a direct and immediate way. That frustration is bad and so a reason to dislike the sensation.

Return to the pain in my ankle. I would like to do many things today—go bushwalking, shop for groceries, play cricket again. My pain weighs against all of these because each of them is incompatible with protecting my ankle. So either I fail to do them and am thereby thwarted, or else I do them but must

1 See, for example, Hall (1989), Hare (1964), and Korsgaard (1996).

push on despite the pain, in which case I am constantly distracted, harried, and pushed away from the smooth pursuit of my goals.[2] Either result is unpleasant. Note too that it is unpleasant for reasons having to do with the sensation itself: it is the *pain* that first and foremost thwarts my plans, not the injury. The injury may or may not prevent my walking, after all (I can hobble around alright). The pain, in contrast, works against my action in a direct way, by motivating protection. That is one of the reasons why I dislike it so much.

Swenson (2009), discussing this aspect of pain's badness, notes that pains are "usurpers" of what he calls "user control." User control is the ability to "consciously and effectively" manipulate some part of your body or some aspect of your internal mental life (208). Typically, I have a great deal of user control: I can choose to move my arm smoothly, to focus on my work, to carry out my plans, and so on.

Pains, Swenson notes, interfere with user control in two characteristic ways. First, they're *invasive*: they work against your plans, even your all-things-considered best ones, and as such feel like an external force working against you.[3] Second, you are made passive with respect to aspects of yourself over which you normally have control. My pain removes the ability to smoothly control my ankle; I must do what *it* wants rather than what *I* want. Similarly, because pains are distracting, I'm made passive with respect to my own psychology: if I can't concentrate, then I must to some degree go along with what's happening to me rather than mastering it (Swenson, 2009).

The same story applies to other intense imperative sensations, and so also partly explains why they are unpleasant. Strong hunger demands that I eat *now*. If I don't want to or can't, then my plans will be similarly frustrated by hunger's insistent demands. Imperative sensations demand actions. Those actions are usually at odds with what we're ordinarily doing just because most of our plans involve doing something other than tending to the basic needs of our body. Note too that if they're *not* at odds, then the sensation may well be pleasant: moderate hunger just before a meal can be pleasant precisely because what hunger demands coincides with our considered plans.

2 Here I think there might be something to the Aristotelian idea that pain is the result of a misfit between activity and capacity (Strohl, 2012). I think this is implausible as a theory of *pain*, but *painfulness* often seems to involve this sort of mismatch with the world.

3 Compare Bakan (1968): "Phenomenally, however, pain appears to the conscious ego as not a part of itself, but as alien to it, as something happening *to* the ego, with the ego, as it were, the victim of external forces" (74).

That said, pain is the paradigmatically frustrating sensation. Hunger rarely demands *immediate* satisfaction; mild hunger can often be put off. Satisfying my hunger within the next few hours is almost always something I can fit into my plans, so hunger doesn't become painful until it gets really urgent. Pains, by contrast, almost inevitably frustrate our plans. That's because pains typically demand cessation of bodily movement in the service of protecting a body part. But my body is what I do things with. So any cessation of movement directly impinges on many things I might be planning on doing.

The link between painfulness and frustrated desires might also explain some of the painfulness of painful emotions. In grief, heartbreak, and wrath, I want something that I can't get—the presence of a loved one, the intimacy of a former beloved, or revenge against a slight. The world conspires such that I cannot have what I want, and if that mismatch is severe enough, the emotion can be painful. Note that this requires the *emotion* to be disliked precisely because it is driving you toward something that cannot be achieved—my wrath drives me toward revenge, my heartbreak toward impossible reunion—and as such it's a reason to dislike the emotion itself.

Finally, the frustration of pain and the like may lead to a second-order problem. Because we dislike pains, we want to get rid of them. Often, we can't. That leads to *further* frustrated desire: now it's not just my ordinary plans that are thwarted but also my desire to be rid of injury in order to carry out my plans. In the case of the painful emotions, I might similarly rail against both my experience and against the world—the laws that prevent revenge, the injustice that took a friend too soon. But this fight is also doomed to failure. My foot will not heal any faster just because I am angry at it. So although we fight painful sensations and their causes, that fighting makes painfulness worse. As Korsgaard (1996) notes,

> If the painfulness of pain rested in the character of the sensations rather than in our tendency to revolt against them, our belief that physical pain has something in common with grief, rage and disappointment would be inexplicable.... What emotional pains have in common with physical ones is that in these cases too we are in the grip of an overwhelming urge to do battle, not now against our sensations, but against the world. (148)

I do not think that this battle against the world explains everything about why we dislike pains. I do think it explains a lot and fits especially well with the badness of chronic and unsatisfiable pains. The badness of chronic pain is not simply the sum of the badnesses of an equal number of acute pains. Chronic pain brings with it its own kind of struggle and with that a new and unique sort of suffering.

14.3.3 The Dual Nature of Pain

Pain has a curious sort of dual nature. Pains are undoubtedly our own sensations. Yet precisely because they can be distracting, unpleasant, and motivating against our otherwise considered aims, they can feel like distinct things opposed to us. If I am hiking, the pain in my ankle is something that thwarts my plans, that makes me suffer, and against which I may rail. In an excellent passage, David Sussman (2005) writes,

> Pain resembles a kind of primitive language of bodily commands and pleas that makes the same kind of insistent demands on our attention and response as our children's shrieking and whining. Understood as a kind of expressive voice, my pain is not unproblematically an exercise of my own agency (the way my reflectively adopted commitments might be), but neither is it something fully distinct from such agency. (21)

The structure of imperative motivation outlined in chapter 6 goes some way toward explaining the curious dual nature of pains. Commands issuing from practical authorities have a similar dual nature. As Sussman (2005) reflects,

> If someone tells me to "Go away!" I cannot first contemplate the utterance and then consider how I might respond to it. Rather, in hearing the command, the possibility of going away has already become real for me, as something I must make some effort to disobey or disregard. I may challenge or ignore the order, but this is always a "second move" in response to an initial proposal in which I find myself to have already begun to participate. (21)

On the one hand, I accept the commands of an authority as reasons for action. That depends on facts about me. Further, the reasons I gain are *my* reasons for action, same as any other. On the other hand, the second-order reasons that I get from practical authorities are quite unlike most of the reasons I have. Authorities not only motivate me to take a certain course of action but also bind my deliberation with respect to those actions. Although I'm the one who's motivated, and motivated by first-order reasons that I have, I am not as free with respect to that motivation as I am with other sorts of reasons.

This is, I suspect, why ordinary practical authorities are so often resented. If an authority commands me to do something I already had good reason to do, then there's a sense in which I don't lose out. But my deliberation might end up bound in a way that it wasn't before, and *that* aspect of deliberation makes accepting authority hard to swallow.

So too with pains. Even if I intended to stop my walk, my pain changes my motivational landscape. I might end up with the same first-order reasons, but my pain also gives me a second-order reason that I didn't have before: it makes my first-order reason mandatory, whereas before it my reasons to stop might have been optional. Hence, bodily imperatives give us reasons that are importantly disanalogous with those that stem from our own desires. Perhaps I do not have control over what I desire, but I do often have second-order reasons that preclude me from considering certain reasons in my deliberations. Further, those second-order reasons are often under my control: cultivating my moral sense, for example, involves cultivating second-order reasons that keep me from considering certain kinds of immoral first-order reasons.

Bodily imperatives, on the other hand, are not optional features of my deliberation. Because I treat my body as a minimal practical authority, I cannot simply exclude its edicts from my practical reasoning. That is why bodily commands at once belong to us and feel like an outside imposition.

The dual nature of pain also explains why pains are typically attention-grabbing. The reasons for action they give us often won't accord with other reasons we have. That gives rise to a practical conflict. I am bushwalking. I feel a pain in my ankle, which tells me to stop. Now I have to decide what to do: I have conflicting reasons for action that I didn't have before. Further, stopping will preclude continuing to walk (at least for a while). So now I must weigh the different reasons I have for action and come up with a new plan. The mandatory nature of pains, further, means I can't avoid conflict just by ignoring my pain. Resolving such practical conflicts thus takes a bit of deliberation, and that deliberation takes attention (Mole, 2011). Hence, pains are attention-grabbing. Indeed, they grab attention over and over in many cases: each fluctuation in the strength of the imperative (let's suppose) gives rise to a reason with a slightly different strength. That requires a bit of recalibration and again more attention. Conversely, deliberation will draw our attention to standing imperatives because we take them to provide reasons for action and so must accommodate them.

14.4 Pain, Fragility, and Mortality

The dual nature of pains is the key to the final, and I think deepest, reason why pains are disliked, hurt, and feel bad.

On the one hand, pains thwart our plans, frustrate us, annoy us, and otherwise work against us. On the other hand, my pain is unmistakably *my* sensation, coming from *me*, issued on behalf of *my* body for reasons that *I* accept. David Sussman (2005) again puts it well:

> Physical pain has a peculiar quality. On the one hand, we experience it as not a part of ourselves: it is something unbidden whose very nature is such that we want to expel it from of [sic] ourselves, to abolish it or drive it away. In an obvious way, we are passive before physical pain; it is something that just happens to us, neither immediately evoked nor eliminated by any decisions or judgments we may make.
>
> On the other hand, pain is also a primitive, unmediated aspect of our own agency. Pain is not something wholly alien to our wills, but something in which we find ourselves actively, if reluctantly, participating. My pain is, after all, my pain, an experience that in its very nature seems to demand, wheedle, or plead with me. (19–20)

So pain is not like just any frustrating external condition. In railing against it, I rail against myself. Swenson's aspects of invasion and passivity might apply to many different bad things. But pain is special: with pain, it is *I* who invade myself. It is *I* who undermine my own control. Elaine Scarry (1985) writes,

> The ceaseless, self-announcing signal of the body in pain, at once so empty and undifferentiated and so full of blaring adversity, contains not only the feeling "my body hurts" but the feeling "my body hurts me." (47)

Because of this "double experience of agency," Scarry continues, pain is not simply one more thing against which we may fight:

> While pain is in part a profound sensory rendering of "against," it is also a rendering of the "something" that is against, a something at once internal and external. Even when there is an actual weapon present, the sufferer may be dominated by a sense of internal agency: it has often been observed that when a knife or a nail or pin enters the body, one feels not the knife, nail or pin but one's own body, one's own body hurting one. Conversely, in the utter absence of any actual external cause, there often arises a vivid sense of external agency, a sense apparent in our elementary, everyday vocabulary for pain: knifelike pains, stabbing, boring, searing pains. In physical pain, then, suicide

and murder converge, for one feels acted upon, annihilated, by inside and outside alike. (52–53)

There can be little doubt that this sort of self-opposition is uniquely unpleasant. Sussman, for example, argues that torture is distinctively wrong precisely because it turns the victim's body—and hence part of his own agency—against himself.[4]

I think imperativism sheds light on this final, deep aspect of pain's unpleasantness. Imperativism is designed to capture the dual nature of pains. As sensations, pains undoubtedly are my own and belong to me. Further, they are issued by my body; so they come from me, in some important sense, rather than from the world. Yet in an important sense, pains are not something I do but are something that are done to me. The body commands me. Barring the sorts of drastic problems discussed in previous chapters, I have a general attitude of acceptance of the body's commands.

However, individual pains are rarely integrated into my plans and desires. Although they aren't integrated, accepting the body as a minimal practical authority means that I still must accept its commands as mandatory in my deliberations. Hence, I'm in a bind: I'm regularly commanded by a source that I know I must accept and yet often seems to undermine my plans.

The attitude of acceptance toward the body presents us with a deep dilemma. For why not simply ignore the body? Even if it's not possible, it's surely a common fantasy: Korsgaard (1996) claims,

> Stoics and Buddhists are right in thinking that we could put an end to pain if we could just stop fighting. The person who cared only for his own virtue, if there could be such a person, would be happy on the rack. (148)

Grahek (2007) has a telling discussion of a system developed by Paul Brand meant to provide stand-in feedback for pain to patients suffering from peripheral numbness as a consequence of Hansen's disease.[5] The system registered

4 Sussman (2005) also notes that any of what I've called the imperative sensations can be, and have been, used for this purpose: "Any suitably intense and relentless craving, whether for food, drugs, sleep, or just quiet, could be the medium through which a suitably constrained and dependent person might be tortured" (27).

5 Hansen's disease causes peripheral neuropathy. The false perception that the disease causes body parts to fall off is a misunderstanding: peripheral neuropathy leads to numbness, which leads to accumulated unhealed injury and ultimately tissue death.

force sufficient to cause injury and shocked the wearer in the armpit as a warning. Wearers simply turned off the machine when it became bothersome. Maladaptive, to be sure, but a familiar feeling: in many cases, if we could simply ignore pain, we could.

But we can't. Why can't we ignore our bodies? Because we depend on our bodies. They keep us alive. We act with our bodies in the world. Without them, we're stuffed. So we have good reason to treat our bodies as a minimal practical authority. But that doesn't mean we have to *like* it. Quite the opposite, in fact.

Pain, more than any other sensation, makes vivid the fragility of my body. In doing so, it makes evident that I am a limited, finite being. Many authors have emphasized that pain represents an assault on the self. Pain does more than that, however. For one, pain makes clear that I *am* a self, contingently carved off from the world. Perhaps other beings can exist as pure thought; pain makes clear to me that *I* can't, that who I am is bound up with my body in a tight and unfortunately non-negotiable way. As Kundera (2001) poetically puts it,

> I think, therefore I am is the statement of an intellectual who underrates toothaches. I feel, therefore I am is a truth much more universally valid, and it applies to everything that's alive.... The basis of the self is not thought but suffering. (chapter 11)

In assaulting the self, pain also makes vivid that I am the sort of thing with a self that can be assaulted. I care about my self-preservation, and thus I *must* care about my body. Every time I feel pain and respect its commands, I am implicitly acknowledging what sort of thing I am: fragile, finite, and ultimately mortal.[6] That is something that nobody likes to be reminded of. That would be sufficient to explain why pains are unpleasant and why we dislike them so vividly.

The imperativist claims that pain is a command, part of an ancient language that the body goads us with to keep us intact. It is, in itself, no worse than hunger, thirst, or any of the other myriad commands that the body issues. We are not merely embodied, however: we have goals and desires that outstrip immediate bodily maintenance. It is through conflict with those goals and desires—those very things that constitute us—that the commands of the body cause us to suffer.

6 I am partly inspired here by Bakan's (1968) discussion of pain as the "psychic manifestation of telic decentralization" (58). By this, I take it, Bakan means that painfulness (in my formulation) partly consists in realizing that one is a composite organism that is dependent on the smooth interworkings of numerous disparate parts, some of which happen at that time not to be working well together and thus threatening the whole system.

Bibliography

Adolph, E. F. (1941). The internal environment and behavior Part III: Water content. *American Journal of Psychiatry*, 97(6):1365–1373.

American Psychiatric Association. (2013). Paraphilic disorders fact sheet.

Anelli, F., Ranzini, M., Nicoletti, R., and Borghi, A. M. (2013). Perceiving object dangerousness: An escape from pain? *Experimental brain research*, 228(4):457–466.

American Psychological Association. (2000). *Diagnostic and statistical manual of mental disorders: DSM–IV–TR.* American Psychological Association, Washington, DC.

Armstrong, D. (1962). *Bodily sensations*. Routledge & Paul, New York.

Aydede, M. (2001). Naturalism, introspection, and direct realism about pain. *Consciousness & Emotion*, 2(1):29–73.

Aydede, M. (2006a). The main difficulty with pain. In Aydede (2006b), pages 123–136.

Aydede, M. (2006b). *Pain: New essays on the nature of pain and the methodology of its study.* MIT Press, Cambridge.

Aydede, M. and Güzeldere, G. (2002). Some foundational problems in the scientific study of pain. *Philosophy of Science*, 69(S3):S265–S283.

Baier, K. (1962). Pains. *Australasian Journal of Philosophy*, 40(1):1–23.

Bain, D. (2011). The imperative view of pain. *Journal of Consciousness Studies*, 18(9–10):164–185.

Bain, D. (2013a). Pains that don't hurt. *Australasian Journal of Philosophy*, 92(2):1–16.

Bain, D. (2013b). What makes pains unpleasant? *Philosophical Studies*, 166(1):69–89.

Bakan, D. (1968). *Disease, pain, and sacrifice: Toward a psychology of suffering.* University of Chicago Press, Chicago.

Barber, T. X. (1959). Toward a theory of pain: Relief of chronic pain by prefrontal leucotomy, opiates, placebos, and hypnosis. *Psychological Bulletin*, 56(6):430–460.

Barnier, A. J., Dienes, Z., and Mitchell, C. J. (2008). How hypnosis happens: New cognitive theories of hypnotic responding. In Nash, M. R. and Barnier, A. J., editors, *The Oxford handbook of hypnosis: Theory, research, and practice*, chapter 6, pages 141–177. Oxford University Press, Oxford.

Barrett, L. (2006). Are emotions natural kinds? *Perspectives on Psychological Science*, 1(1):28–58.

Baumeister, R. F. (1989). *Masochism and the self.* Lawrence Erlbaum Associates, Hillsdale.

Bayne, T. (2010a). Agentive experiences as pushmi-pullyu representations. In *New Waves in the Philosophy of Action*, pages 219–236. Palgrave Macmillan, New York.

Bayne, T. (2010b). *The unity of consciousness*. Oxford University Press, New York.

Bearn, G. (2013). *Life drawing: A Deleuzean aesthetics of existence*. Fordham University Press, New York.

Beecher, H. (1956). Relationship of significance of wound to pain experienced. *Journal of the American Medical Association*, CLXI:1609–1613.

Bentley, T. (2006). *The surrender*. HarperCollins UK, London.

Berthier, M., Starkstein, S., and Leiguarda, R. (1988). Asymbolia for pain: A sensory-limbic disconnection syndrome. *Annals of Neurology*, 24(1):41–49.

Block, N. (2006). Bodily sensations as an obstacle for representationism. In Aydede (2006b), pages 137–142.

Bonnot, O., Anderson, G., Cohen, D., Willer, J., and Tordjman, S. (2009). Are patients with schizophrenia insensitive to pain? A reconsideration of the question. *The Clinical Journal of Pain*, 25(3):244–252.

Brang, D., McGeoch, P. D., and Ramachandran, V. S. (2008). Apotemnophilia: A neurological disorder. *Neuroreport*, 19(13):1305–1306.

Bray, H. and Moseley, G. L. (2011). Disrupted working body schema of the trunk in people with back pain. *British Journal of Sports Medicine*, 45(3):168–173.

Burroughs, W. (1957). Letter from a master addict to dangerous drugs. *British Journal of Addiction to Alcohol & Other Drugs*, 53(2):119–132.

Byrne, A. (2001). Intentionalism defended. *The Philosophical Review*, 110(2):199–240.

Byrne, A. and Hilbert, D. R. (2003). Color realism and color science. *Behavioral and Brain Sciences*, 26(1):3–21.

Cannon, W. B. (1932). *The wisdom of the body.* W.W. Norton & Co, New York.

Catani, M. et al. (2005). The rises and falls of disconnection syndromes. *Brain*, 128(10):2224–2239.

Chemero, A. (2003). An outline of a theory of affordances. *Ecological Psychology*, 15(2):181–195.

Chemero, A. (2009). *Radical embodied cognitive science.* The MIT Press, Cambridge.

Clark, A. (1995). Moving minds: Situating content in the service of real-time success. *Philosophical Perspectives*, pages 89–104.

Clark, J. (1972). *Ignition! An informal history of liquid rocket propellants.* Rutgers University Press, New Brunswick.

Coco, A. S. (1999). Primary dysmenorrhea. *American Family Physician*, 60(2):489–496.

Cooper, J. M. (1996). An Aristotleian theory of the emotions. In Rorty, A., editor, *Essays on Aristotle's Rhetoric*, pages 238–257. University of California Press, Berkeley.

Craig, A. (2002). How do you feel? Interoception: The sense of the physiological condition of the body. *Nature Reviews Neuroscience*, 3:655–666.

Craig, A. (2003a). A new view of pain as a homeostatic emotion. *Trends in Neurosciences*, 26(6):303–307.

Craig, A. (2003b). Pain mechanisms: Labeled lines versus convergence in central processing. *Annual Review of Neuroscience*, 26(1):1–30.

Craig, A. (2008). Interoception and emotion: A neuroanatomical perspective. In Lewis, M., Haviland-Jones, J. M., and Barrett, L. F., editors, *Handbook of emotion*, pages 272–288. The Guilford Press, New York.

Crawford, C. S. (2009). From pleasure to pain: The role of the MPQ in the language of phantom limb pain. *Social Science & Medicine*, 69:655–661.

Cutter, B. and Tye, M. (2011). Tracking representationalism and the painfulness of pain. *Philosophical Issues*, 21(1):90–109.

Darwin, E. (1918). *Zoonomia*, volume 1. Edward Earle, Philadelphia.

de Vignemont, F. (2010). Body schema and body image–pros and cons. *Neuropsychologia*, 48(3):669–680.

Deligeoroglou, E. (2000). Dysmenorrhea. *Annals of the New York Academy of Sciences*, 900(1):237–244.

Dennett, D. C. (1985). Why you can't make a computer that feels pain. In *Brainstorm: Philosophical essays on mind and psychology*, pages 190–232. MIT Press, Cambridge.

Denton, D. (2006). *The Primordial emotions.* Oxford University Press, New York.

Descartes, R. (2006). *Meditations on first philosophy.* R. Ariew & D. Cress, Trans. Hackett Publishing Company, Indianapolis.

Dutton, D. and Aron, A. (1974). Some evidence for heightened sexual attraction under conditions of high anxiety. *Journal of Personality and Social Psychology*, 30(4):510–517.

Dworkin, R. (1994). Pain insensitivity in schizophrenia. *Schizophrenia Bulletin*, 20(2):235–248.

Egan, A. (2006). Appearance properties? *Noûs*, 40(3):495–521.

Eisenberger, N. I. and Lieberman, M. D. (2004). Why rejection hurts: A common neural alarm system for physical and social pain. *Trends in Cognitive Sciences*, 8(7):294–300.

Ellsberg, D. (1954). Classic and current notions of "measurable utility". *The Economic Journal*, 64(255):528–556.

Elster, J. (2009). Urgency. *Inquiry*, 52(4):399–411.

Feldman, F. (2004). *Pleasure and the good life: Concerning the nature, varieties and plausibility of hedonism*. Clarendon Press, Oxford.

Fernandez, E. and Turk, D. C. (1992). Sensory and affective components of pain: Separation and synthesis. *Psychological Bulletin*, 112(2):205–217.

Fowler, C. J., Griffiths, D., and de Groat, W. C. (2008). The neural control of micturition. *Nature Reviews Neuroscience*, 9(6):453–466.

Gallagher, S. (1986). Body image and body schema: A conceptual clarification. *Journal of Mind and Behavior*, 7(4):541–554.

Gallagher, S. (2005). *How the body shapes the mind*. Cambridge University Press, Cambridge.

Gallagher, S., Butterworth, G. E., Lew, A., and Cole, J. (1998). Hand–mouth coordination, congenital absence of limb, and evidence for innate body schemas. *Brain and Cognition*, 38(1): 53–65.

Ganson, T. and Bronner, B. (2013). Visual prominence and representationalism. *Philosophical Studies*, 164(2):405–418.

Gawande, A. (2008). The itch. *The New Yorker*, 6(30):58–65.

Geschwind, N. (1965). Disconnexion syndromes in animals and man. *Brain*, 88:237–294.

Gracely, R. H., McGrath, P., and Dubner, R. (1978). Ratio scales of sensory and affective verbal pain descriptors. *Pain*, 5(1):5–18.

Grahek, N. (2007). *Feeling pain and being in pain*. MIT Press, Cambridge, 2nd edition.

Graziano, M.S. (2006). The organization of behavioral repertoire in motor cortex. *Annual Review of Neurosciences*, 29:105–134.

Graziano, M. S. and Aflalo, T. N. (2007). Rethinking cortical organization: Moving away from discrete areas arranged in hierarchies. *The Neuroscientist*, 13(2):138–147.

Graziano, M. S. and Cooke, D. F. (2006). Parieto-frontal interactions, personal space, and defensive behavior. *Neuropsychologia*, 44(6):845–859.

Graziano, M. S., Taylor, C. S., and Moore, T. (2002a). Complex movements evoked by microstimulation of precentral cortex. *Neuron*, 34(5):841–851.

Graziano, M. S., Taylor, C. S., Moore, T., and Cooke, D. F. (2002b). The cortical control of movement revisited. *Neuron*, 36(3):349–362.

Griffiths, P. (1997). *What emotions really are: The problem of psychological categories*. University of Chicago Press, Chicago.

Guieu, R., Samuelian, J., and Coulouvrat, H. (1994). Objective evaluation of pain perception in patients with schizophrenia. *The British Journal of Psychiatry*, 164(2):253.

Hall, R. (1989). Are pains necessarily unpleasant? *Philosophy and Phenomenological Research*, 49(4):643–659.

Hall, R. (2008). If it itches, scratch! *Australasian Journal of Philosophy*, 86(4):525–535.

Hall, W. (1981). On "Ratio scales of sensory and affective verbal pain descriptors." *Pain*, 11(1):101–106.

Hamblin, C.L. (1987). *Imperatives*. Basil Blackwell, Oxford.

Hardcastle, V. (1997). When a pain is not. *The Journal of Philosophy*, 94(8):381–409.

Hardcastle, V. G. (1999). *The myth of pain*. MIT Press, Cambridge.

Hare, R. M. (1964). Symposium: Pain and evil. *Proceedings of the Aristotelian Society, Supplementary Volumes*, 38:91–124.

Harman, G. (1990). The intrinsic quality of experience. *Philosophical Perspectives*, 4:31–52.

Hart, H. L. A. (1982). *Essays on Bentham: Studies in jurisprudence and political theory*. Clarendon Press, Oxford.

Helm, B. (2002). Felt evaluations: A theory of pleasure and pain. *American Philosophical Quarterly*, 39(1):13–30.

Hemphill, R. and Stengel, E. (1940). A study on pure word-deafness. *Journal of Neurology and Psychiatry*, 3(3):251–262.

Hilbert, D. R. (1987). *Color and color perception: A study in anthropocentric realism*. Center for the Study of Language and Information, Stanford.

Hobbes, T. (1651). *Leviathan*. Penguin Classics, New York.

Huttegger, S. M. (2007). Evolutionary explanations of indicatives and imperatives. *Erkenntnis*, 66(3):409–436.

IASP Task Force on Taxonomy. (1994). Part III: Pain terms, a current list with definitions and notes on usage. In Merksey, H. and Bogduk, N., editors, *Classification of chronic pain*, pages 209–214. International Association for the Study of Pain Press, Seattle, 2nd edition.

Ibañez, A., Gleichgerrcht, E., and Manes, F. (2010). Clinical effects of insular damage in humans. *Brain Structure and Function*, 214:1–14.

Kaas, J. H., Gharbawie, O. A., and Stepniewska, I. (2013). Cortical networks for ethologically relevant behaviors in primates. *American Journal of Primatology*, 75(5):407–414.

Kant, I. (1912). *The critique of judgment*. Macmillan and Co., London.

Karnath, H. and Baier, B. (2010). Right insula for our sense of limb ownership and self-awareness of actions. *Brain Structure and Function*, 214(5):411–417.

Keats, A. and Beecher, H. (1950). Pain relief with hypnotic doses of barbiturates and a hypothesis. *Journal of Pharmacology and Experimental Therapeutics*, 100(1):1–13.

Keegan, J. (2011). *The face of battle: A study of Agincourt, Waterloo and the Somme*. Random House, Sydney.

Klein, C. (2007). An imperative theory of pain. *Journal of Philosophy*, 104(10):517–532.

Klein, C. (2010). Response to Tumulty on pain and imperatives. *The Journal of Philosophy*, CVII(10):554–557.

Klein, C. (2012). Imperatives, phantom pains, and hallucination by presupposition. *Philosophical Psychology*, 25(6):917–928.

Klein, C. (2014). The penumbral theory of masochistic pleasure. *Review of Philosophy and Psychology*, 5(1):41–55.

Korsgaard, C. (1996). *The sources of normativity*. Cambridge University Press, Cambridge.

Kripke, S. (1981). *Naming and necessity*. Wiley-Blackwell, New York.

Kundera, M. (2001). *Immortality*. Faber & Faber, Limited, New York.

Lewis, D. (1969). *Convention: A philosophical study*. Harvard University Press, Cambridge.

Lewis, J. S., Kersten, P., McCabe, C. S., McPherson, K. M., and Blake, D. R. (2007). Body perception disturbance: A contribution to pain in complex regional pain syndrome (crps). *Pain*, 133(1):111–119.

Liang, C. and Lane, T. (2009). Higher-order thought and pathological self: The case of somatoparaphrenia. *Analysis*, 61:661–668.

MacDonald, G. and Leary, M. R. (2005). Why does social exclusion hurt? The relationship between social and physical pain. *Psychological Bulletin*, 131(2):202–223.

Martínez, M. (2010). Imperative content and the painfulness of pain. *Phenomenology and the Cognitive Sciences*, 10:67–90.

Matthen, M. P. (2005). *Seeing, doing, and knowing: A philosophical theory of sense perception*. Oxford University Press, New York.

Meduna, L. J. (1950). *Carbon dioxide therapy: A neurophysiological treatment of nervous disorders*. Charles C Thomas, Springfield.

Melzack, R. (1973). *The puzzle of pain*. Basic Books, New York.

Melzack, R. (1983). The McGill pain questionnaire. In *Pain measurement and assessment*, pages 41–47. Raven Press, New York.

Melzack, R. (1984). The myth of painless childbirth (The John J. Bonica lecture). *Pain*, 19(4):321–337.

Melzack, R. (1990). Phantom limbs and the concept of a neuromatrix. *Trends in Neurosciences*, 13(3):88–92.

Melzack, R. (1999). From the gate to the neuromatrix. *Pain*, 82:S121–S126.

Melzack, R. and Katz, J. (2006). Pain assessment in adult patients. In McMahon, S. B. and Koltzenburg, M., editors, *Wall and Melzack's textbook of pain*, pages 291–304. Elsevier, London, 5th edition.

Melzack, R. and Wall, P. (1965). Pain mechanisms: A new theory. *Science*, 150(699):971–979.

Melzack, R. and Wall, P. (1996). *The challenge of pain*. Penguin, New York.

Melzack, R., Wall, P., and Ty, T. (1982). Acute pain in an emergency clinic: Latency of onset and descriptor patterns related to different injuries. *Pain*, 14(1):33–43.

Menzies, P. (2007). Causation in context. In Corry, R. and Price, H., editors, *Causation, physics, and the constitution of reality: Russell's Republic Revisited*, pages 191–223. Oxford University Press, Oxford.

Millan, M. J. (2002). Descending control of pain. *Progress in Neurobiology*, 66(6):355–474.

Millikan, R. G. (1995). Pushmi-pullyu representations. *Philosophical Perspectives*, IX:185–200.

Millikan, R. G. (2004). *Varieties of meaning: The 2002 Jean Nicod lectures*. The MIT Press, Cambridge.

Minsky, M. (1988). *Society of mind*. Simon and Schuster, New York.

Mole, C. (2011). *Attention is cognitive unison: An essay in philosophical psychology*. Oxford University Press, New York.

Montgomery, G. H., DuHamel, K. N., and Redd, W. H. (2000). A meta-analysis of hypnotically induced analgesia: How effective is hypnosis? *The International Journal of Clinical and Experimental Hypnosis*, 48(2):138–153.

Moseley, G. L. (2008). I cant find it! Distorted body image and tactile dysfunction in patients with chronic back pain. *Pain*, 140(1):239–243.

Nanay, B. (2011). Do we see apples as edible? *Pacific Philosophical Quarterly*, 92(3):305–322.

Nash, T. (2005). Editorial II: What use is pain? *British Journal of Anaesthesia*, 94(2):146–149.

Neander, K. (1995). Misrepresenting & malfunctioning. *Philosophical Studies*, 79(2):109–141.

Nelkin, N. (1986). Pains and Pain sensations. *The Journal of Philosophy*, 83(3):129–148.

Northoff, G. (2002). What catatonia can tell us about "top-down modulation": A neuropsychiatric hypothesis. *Behavioral and Brain Sciences*, 25(5):555–577.

Noyes, R. and Kletti, R. (1977). Depersonalization in response to life-threatening danger. *Comprehensive Psychiatry*, 18(4):375–384.

Parsons, J. (2012a). Cognitivism about imperatives. *Analysis*, 72(1):49–54.

Parsons, J. (2012b). Command and consequence. *Philosophical Studies*, 164:1–32.

Patterson, D. R. (2004). Treating pain with hypnosis. *Current Directions in Psychological Science*, 13(6):252–255.

Patterson, D. R. and Jensen, M. P. (2003). Hypnosis and clinical pain. *Psychological Bulletin*, 129(4):495–521.

Perl, E. R. (2007). Ideas about pain, a historical view. *Nature Reviews Neuroscience*, 8(1):71–80.

Pitcher, G. (1970). The awfulness of pain. *The Journal of Philosophy*, 67(14):481–492.

Ploner, M., Freund, H., and Schnitzler, A. (1999). Pain affect without pain sensation in a patient with a postcentral lesion. *Pain*, 81(1-2):211–214.

Plum, F. and Posner, J. B. (2007). *The diagnosis of stupor & coma*. Oxford University Press, Oxford, 4th edition.

Portner, P. (2007). Imperatives and modals. *Natural Language Semantics*, 15(4):351–383.

Price, D. (2000). Psychological and neural mechanisms of the affective dimension of pain. *Science*, 288(5472):1769–1772.

Price, D. D. and Aydede, M. (2006). The experimental use of introspection in the scientific study of pain and its integration with third-person methodologies: The experiential-phenomenological approach. In Aydede (2006b), pages 243–274.

Price, D. D., Barrell, J. J., and Rainville, P. (2002). Integrating experiential–phenomenological methods and neuroscience to study neural mechanisms of pain and consciousness. *Consciousness and Cognition*, 11(4):593–608.

Price, D. D. and Dubner, R. (1977). Mechanisms of first and second pain in the peripheral and central nervous systems. *Journal of Investigative Dermatology*, 69(1):167–171.

Price, D. D., McGrath, P. A., Rafii, A., and Buckingham, B. (1983). The validation of visual analogue scales as ratio scale measures for chronic and experimental pain. *Pain*, 17:45–56.

Price, J. L. (2005). Free will versus survival: Brain systems that underlie intrinsic constraints on behavior. *Journal of Comparative Neurology*, 493(1):132–139.

Rainville, P., Carrier, B. T., Hofbauer, R. K., Bushnell, M. C., and Duncan, G. H. (1999). Dissociation of sensory and affective dimensions of pain using hypnotic modulation. *Pain*, 82(2):159–171.

Rainville, P., Duncan, G., Price, D., Carrier, B., and Bushnell, M. (1997). Pain affect encoded in human anterior cingulate but not somatosensory cortex. *Science*, 277:968–971.

Ramachandran, V.S. (1998). Consciousness and body image: Lessons from phantom limbs, Capgras syndrome and pain asymbolia. *Philosophical Transactions of the Royal Society B: Biological Sciences*, 353(1377):1851–1859.

Ramachandran, V.S. and Blakeslee, S. (1999). *Phantoms in the brain: Probing the mysteries of the human mind*. William Norton Co., New York.

Ramachandran, V.S. and McGeoch, P. (2007). Occurrence of phantom genitalia after gender reassignment surgery. *Medical Hypotheses*, 69(5):1001–1003.

Ramachandran, V. S. and Hirstein, W. (1998). The perception of phantom limbs (the DO Hebb Lecture). *Brain*, 121(9):1603–1630.

Ramachandran, V. S., McGeoch, P. D., Williams, L., and Arcilla, G. (2007). Rapid relief of thalamic pain syndrome induced by vestibular caloric stimulation. *Neurocase*, 13(3):185–188.

Raz, J. (1975). Reasons for action, decisions and norms. *Mind*, 84(1):481–499.

Raz, J. (1986). *The morality of freedom*. Oxford University Press, Oxford.

Raz, J. (1999). *Practical reason and norms*. Oxford University Press, New York.

Redelmeier, D. A., Katz, J., and Kahneman, D. (2003). Memories of colonoscopy: A randomized trial. *Pain*, 104(1):187–194.

Reed, E. (1996). *Encountering the world: Toward an ecological psychology*. Oxford University Press, New York.

Riley III, J. L., Robinson, M. E., Wise, E. A., and Price, D. D. (1999). A meta-analytic review of pain perception across the menstrual cycle. *Pain*, 81:225–235.

Röder, C., Michal, M., Overbeck, G., van de Ven, V., and Linden, D. (2000). Pain response in depersonalization: A functional imaging study using hypnosis in healthy subjects. *Psychotherapy and Psychosomatics*, 76:115–121.

Rosenthal, D. (2005). *Consciousness and mind*. Oxford University Press, New York.

Ross, A. (1944). Imperatives and logic. *Philosophy of Science*, 11(1):30–46.

Sacks, O. (1986). *Migraine: Understanding a common disorder (expanded and updated)*. University of California Press, Berkeley.

Scarry, E. (1985). *The body in pain: The making and unmaking of the world*. Oxford University Press, New York.

Schachter, S. and Singer, J. (1962). Cognitive, social and physiological determinants of emotional state. *Psychological Review*, 69(5):379–399.

Schilder, P. and Stengel, E. (1928). Schmerzasymbolie. *Zeitschrift für die gesamte Neurologie und Psychiatrie*, 113(1):143–158.

Schilder, P. and Stengel, E. (1931). Asymbolia for pain. *Archives of Neurology & Psychiatry*, 25(3):598–600.

Schmidt, J. O., Blum, M. S., and Overal, W. L. (1984). Hemolytic activities of stinging insect venoms. *Archives of Insect Biochemistry and Physiology*, 1(2):155–160.

Schott, J. and Rossor, M. (2003). The grasp and other primitive reflexes. *Journal of Neurology, Neurosurgery & Psychiatry*, 74(5):558–560.

Schroeder, M. (2008). *How* does the good appear to us? *Social Theory and Practice*, 34(1):119–130.

Schwoebel, J. and Coslett, H. B. (2005). Evidence for multiple, distinct representations of the human body. *Journal of Cognitive Neuroscience*, 17(4):543–553.

Schwoebel, J., Friedman, R., Duda, N., and Coslett, H. B. (2001). Pain and the body schema evidence for peripheral effects on mental representations of movement. *Brain*, 124(10):2098–2104.

Searle, J. R. (1985). *Expression and meaning: Studies in the theory of speech acts*. Cambridge University Press, Cambridge.

Shallice, T. (1988). *From neuropsychology to mental structure*. Cambridge University Press, Cambridge.

Sherrington, C. (1961). *The integrative action of the nervous system*. Yale University Press, New Haven, 2nd edition.

Siegel, S. (2014). Affordances and the contents of perception. In Brogaard, B., editor, *Does perception have content?* Oxford University Press, New York.

Sierra, M. (2009). *Depersonalization: A new look at a neglected syndrome*. Cambridge University Press, New York.

Singer, T., Critchley, H., and Preuschoff, K. (2009). A common role of insula in feelings, empathy and uncertainty. *Trends in Cognitive Sciences*, 13(8):334–340.

Singer, T., Seymour, B., O'Doherty, J., Kaube, H., Dolan, R. J., and Frith, C. D. (2004). Empathy for pain involves the affective but not sensory components of pain. *Science*, 303(5661):1157–1162.

Singh, M., Giles, L., and Nasrallah, H. (2006). Pain insensitivity in schizophrenia: Trait or state marker? *Journal of Psychiatric Practice*, 12(2):90–102.

Skyrms, B. (2010). *Signals: Evolution, learning, and information*. Oxford University Press, Oxford.

Smart, J. (1959). Sensations and brain processes. *The Philosophical Review*, 68(2):141–156.

Smuts, A. (2011). The feels good theory of pleasure. *Philosophical Studies*, 155(2):241–265.

Sprague, J. (1966). Interaction of cortex and superior colliculus in mediation of visually guided behavior in the cat. *Science*, 153(3743):1544–1547.

Starr, C. K. (1985). A simple pain scale for field comparison of hymenopteran stings. *Journal of Entomological Science*, 20(2):225–231.

Sterelny, K. (2003). *Thought in a hostile world: The evolution of human cognition*. Blackwell Publishers, Malden.

Sternbach, R. (1963). Congenital insensitivity to pain: A critique. *Psychological Bulletin*, 60(3):252–264.

Stimmel, B. (1983). *Pain, analgesia, and addiction*. Raven Press, New York.

Stoljar, D. (2004). The argument from diaphanousness. In M. Ezcurdia, R. S. and Viger, C., editors, *New essays in the philosophy of language and mind. Supplemental volume of The Canadian Journal of Philosophy*, pages 341–390. University of Calgary Press, Calgary.

Stoller, R. J. (1991). *Pain & passion: A psychoanalyst explores the world of S & M*. Plenum Press, New York.

Strohl, M. (2012). Horror and hedonic ambivalence. *The Journal of Aesthetics and Art Criticism*, 70(2):203–212.

Sullivan, M., Bishop, S., and Pivik, J. (1995). The pain catastrophizing scale: Development and validation. *Psychological Assessment*, 7(4):524–532.

Sullivan, M. J., Thorn, B., Haythornthwaite, J. A., Keefe, F., Martin, M., Bradley, L. A., and Lefebvre, J. C. (2001). Theoretical perspectives on the relation between catastrophizing and pain. *The Clinical Journal of Pain*, 17(1):52–64.

Sussman, D. (2005). What's wrong with torture? *Philosophy & Public Affairs*, 33(1):1–33.

Swenson, A. (2006). *Pain and value*. PhD thesis, Rutgers.

Swenson, A. (2009). Pain's evils. *Utilitas*, 21(2):197–216.

Taormino, T., editor. (2012). *The ultimate guide to kink: BDSM, role play, and the erotic edge*. Cleis Press, Berkeley.

Tumulty, M. (2009). Pains, imperatives, and intentionalism. *The Journal of Philosophy*, 106(3):161–166.

Tye, M. (1995). A representational theory of pains and their phenomenal character. *Philosophical Perspectives*, IX:223–239.

Tye, M. (2006). Another look at representationalism about pain. In Aydede (2006b), pages 99–120.

Vlaeyen, J. W. and Linton, S. J. (2000). Fear-avoidance and its consequences in chronic musculoskeletal pain: A state of the art. *Pain*, 85(3):317–332.

Vranas, P. (2008). New foundations for imperative logic I: Logical connectives, consistency, and quantifiers. *Noûs*, 42(4):529–572.

Vranas, P. B. M. (2010). In defense of imperative inference. *Journal of Philosophical Logic*, 39(1):59–71.

Wall, P. (1979a). On the relation of injury to pain (The John J. Bonica Lecture). *Pain*, 6(3):253–264.

Wall, P. (1979b). Three phases of evil: The relation of injury to pain. In *Brain and Mind*, pages 293–304. Ciba Foundation, New York.

Wall, P. (2000). *Pain: The science of suffering*. Columbia University Press, New York.

Wall, P. and McMahon, S. (1985). Microneuronography and its relation to perceived sensation. A critical review. *Pain*, 21:209–229.

Waters, C. K. (2007). Causes that make a difference. *The Journal of Philosophy*, 104(11):551–579.

Wittgenstein, L. (1953). *Philosophical investigations*. Basil Blackwell, Oxford.

Woodward, J. (2003). *Making things happen*. Oxford University Press, New York.

Wylie, K. and Tregellas, J. (2010). The role of the insula in schizophrenia. *Schizophrenia Research*, 123(2-3):93–104.

Young, S. (2004). *Break through pain: A step by step mindfulness meditation program for transforming chronic and acute pain*. Sounds True, Boulder.

Zollman, K. J. (2011). Separating directives and assertions using simple signaling games. *Journal of Philosophy*, 108(3):158–169.

Zuckert, R. (2002). A new look at Kant's theory of pleasure. *Journal of Aesthetics and Art Criticism*, 60(3):239–252.

Index